Daniel O'Reilly

PYTHON FOR BEGINNERS

A Complete Beginner's Guide to learning Python quickly.

© Copyright 2020 - All rights reserved.

The content contained within this book may not be reproduced, duplicated or transmitted without direct written permission from the author or the publisher.

Under no circumstances will any blame or legal responsibility be held against the publisher, or author, for any damages, reparation, or monetary loss due to the information contained within this book. Either directly or indirectly.

Legal Notice:

This book is copyright protected. This book is only for personal use. You cannot amend, distribute, sell, use, quote or paraphrase any part, or the content within this book, without the consent of the author or publisher.

Disclaimer Notice:

Please note the information contained within this document is for educational and entertainment purposes only. All effort has been executed to present accurate, up to date, and reliable, complete information. No warranties of any kind are declared or implied. Readers acknowledge that the author is not engaging in the rendering of legal, financial, medical or professional advice. The content within this book has been derived from various sources. Please consult a licensed professional before attempting any techniques outlined in this book.

By reading this document, the reader agrees that under no circumstances is the author responsible for any losses, direct or indirect, which are incurred as a result of the use of information contained within this document, including, but not limited to, errors, omissions, or inaccuracies.

Table Of Contents

Introduction 6
Chapter 1: Introduction to Python 8
Chapter 2: Which Version Is The Easiest To Use For Beginners? 18
Chapter 3: Basics of Python Programming 24
Chapter 4: Python Data Types 46
Chapter 5: The Python Variables 64
Chapter 6: Basic Syntax 68
Chapter 7: Classes in Python 78
Chapter 8: Conditions and Loops 94
Chapter 9: Strings 116
Chapter 10: Functions 128
Chapter 11: Dictionaries 142
Chapter 12: Python Operators 146
Chapter 13: Working with Files 156
Conclusion 170

Introduction

There are a lot of reasons why you will love working with the Python code. It is easy to use, easy to learn, has a lot of great frameworks and libraries to work with (and we will discuss at least a few of these as we go through this guidebook), and is still powerful enough to make machine learning easy for you.

While it is possible to work with other coding languages to help you get the results that you want, but most people prefer to work with Python due to all of the benefits that we have discussed. Before we take a look at how to set up the Python environment so you are able to use it properly, let's take a look at a few of the different parts that come with the Python language, so you understand how a few of these codes can work for you. Programming is the practice of writing instructions for computers to follow. Back in the day, this was written using machine languages. Now, we have higher level languages that are more readable and can be written for any computer to be able to follow. That's why you can find software that will work on any PC of a certain Operating System, or any PC or any Mac or other computer types and operating systems. here are a lot of reasons why you will love working with the Python code. It is easy to use, easy to learn, has a lot of great frameworks and libraries to work with and is still powerful enough to make machine learning easy for you.

While it is possible to work with other coding languages to help you get the results that you want, but most people prefer to work with Python due to all of the benefits that we have discussed. In the beginning, you can take the code and just paste it in the editor to see the results. In the second phase, you can make minor edits to the code and see the results. In the third phase, you will be able to completely reshape a program and see how it runs in the Python shell. Given the increasing applications of Python, learning it is extremely profitable from the angle of the global job market. Python can give you the much-needed edge over others when it comes to securing high paid jobs. In general, programming is a very exciting field. It covers almost every field of what we fondly and generally refer to as technology. Did you know that at its core, your TV, phone, smart fridge, etc. is a computer program? It is true; almost every bit of tech today has its core operations in a computer program.

Want to program an app for your phone? Want to get a job in programming? This book is for you. Once you learn the basics of any programming language, others will come easily. Every programming language can be broken down into basic concepts such as variables, if-statements, and loops. If you are ready to embrace and learn this coding method, then sit back and relax as we take you on a journey into the world of Python programming.

If you are as excited as I am to learn everything Python and how to program with it, let us get started.

CHAPTER 1:

Introduction to Python

What is Python?

Python is one of the most important programming languages nowadays, being a general-purpose language.

With this language, you can create a huge and varied amount of applications, because it allows you to create different kinds of applications since it doesn't have a defined purpose.

Python is an awesome decision on machine learning for a few reasons. Most importantly, it's a basic dialect at the first glance. Regardless of whether you're not acquainted with Python, getting up to speed is snappy in the event that you at any point have utilized some other dialect with C-like grammar.

Second, Python has an incredible network that results in great documentation and extensive answers in StackOverflow (central!).

Third, coming from the colossal network, there are a lot of valuable libraries for Python (both as "batteries included" an outsider), which take care of essentially any issue that you can have (counting machine learning).

Wait I thought this machine language was slow?

Unfortunately, it is a very valid question that deserves an answer. Indeed, Python is not at all the fastest language on the planet.

However, here's the caveat: libraries can and do offload the costly computations to the substantially more performant (yet much harder to use) C and C++ are prime examples. There's NumPy, which is a library for numerical calculation. It is composed in C, and it's quick. For all intents and purposes, each library out there that includes serious estimations utilizes it; every one of the libraries recorded next utilizes it in some shape. On the off chance that you read NumPy, think quickly. In this way, you can influence your computer scripts to run essentially as quick as handwriting them out in a lower level dialect. So there's truly nothing to stress over with regards to speed and agility. To create programs using this type of programming method, you will have to get an application to enter the code into. There is an alternative to getting an application like this, which is writing the application using what is called machine code. This process is a tough one and may make you frustrated and force you to give up on the task altogether. The applications out there made for Python will allow you to type in the code and run it to see what it does in real-time. This means that you will be able to find and address mistakes with ease when downloading these programs. In this chapter, you will be able to find out how to get these programs and how to use them for your particular needs.

Understanding the Platform

Before you start using these programs, you will need to understand what a platform is. The short explanation of a platform is the combination of the computer's hardware with an operating system. A platform will have specialized rules that dictate how it will run. The details of how a platform runs will be hidden from you in the Python application. This means that as you type in the code, the Python application will turn it into something that can be used by the platform as a layout. To make this program work successfully, it will require you to find a Python version that works for the particular operating system that you are running. Due to the overwhelming number of platforms out there, you will have to invest some time and effort into getting the right Python program. By going here, you will be able to get a list of different downloads. The main part of this page will have links for the most popular downloads like Mac, Windows, or Linux. There is a list of links on the left of the screen that give you alternate Python settings to use if needed. If you are looking for a more updated editor than what comes in the regular download, you will be able to find it in these alternate settings. Regardless of what platform you are using, you will be able to find a Python program that works for it.

How to Install Python

Python is a common programming language for application development. Python design focuses on code readability and clear programming for both small and big projects. You are able to run modules and full applications from a massive library of resources on the server. Python works on various operating systems, such as Windows. Installing Python on the Windows server is a straightforward process of downloading the installer and running it on your server and configuring some adjustments that can make working with Python easier.

To begin working with Python, you will first need permission from the Python interpreter. There are different ways of achieving these:

Python is available from the website Python.org. Download the right installer for your operating system and run it on the computer.

Linux provides a packaging manager that you can use to run your installed Python.

The macOS uses a package manager called Homebrew to install Python.

iOS and Android are mobile operating systems, and you will need to install apps that will support Python programming. And this is the best way to practice your coding skills at any time.

And the other way is by finding websites that provide the users the opportunity to get into a Python interpreter online, and there is no need to install it on their computer. And also a chance that the Python will be preinstalled in your operating system. If that's the case, then the version might be outdated, and you need to get the latest version.

Windows

Python does not come preinstalled in the Windows systems. However, installing Python is a straightforward process. All you need to do is to download the Python installer on the website, Python.org, and then run the program.

Here is how you install it on Windows:

Downloading the Python installer:

- Opening your browser, get into the Python.org website, and move to the download page for Windows.

- Click on the heading "Python release for Windows" and click on the link.

- Scroll down and select Windows 64-bit or 32-bit Windows.

When using Windows, you either download a 64 or a 32-bit. The difference between these two is,

- Use a 32-bit installer if you have a 32-bit processor.

- If you have a 64-bit system, then both of the versions will work on both of the purposes. Using a 32-bit will require less memory; however, using a 64-bit will work best with the applications of intensive computation.

- When you are not sure of which version to install, choose a 64-bit version.

Additionally, when you install the incorrect version, and you wish to change it and choose another version, you may do that by uninstalling Python and reinstalling it by getting the version from Python.org.

Running the Installer

When you chose and downloaded the installer, then double-click the downloaded file and run it. You will see a dialog box that has a prompting button instructing you to run the installation. Don't forget to click the box that indicates "Adding Python to PATH"; this will confirm that the interpreter is added in the execution path. The final step is to click 'Install now,' and this is all the required steps. After a few moments, a working Python will be installed on the system.

Windows Subsystem for Linux

When your computer has Windows 10 Creators installed, you have a different way of installing Python. A feature called Windows Subsystem for Linux is used in this version, and you can run the Linux environment straight into Windows without having to use the virtual machine. After you have installed the type of Linux distribution you require, go ahead and install Python from the console window. A similar way can be done if you are natively using a Linux distribution. Most Linux distribution comes with preinstalled Python; however, it might not be the latest version. To be able to know what version you have, there is a process you can do to find out.

Open a prompt window and then try the following commands below:

Python version.

Python2 version.

Python3 version.

The version of the commands from the three commands will answer with the version. If it is not the latest version, you will install the newest version. The procedure of this depends on the Linux distribution running on your computer. It's much useful to use Pyenv, a tool used for multiple Python version management on Linux.

Ubuntu

There are different versions of the Ubuntu distribution, and the installations are different. Identify your Ubuntu versions by trying these commands:

$lsb release a

No-LSB module available

Distributors ID: Ubuntu

Descriptions: Ubuntu-16.04.4.

Release-16.04.

Code name-xenial.

Ubuntu versions 16.10 and 17.04 don't have the Python 3.6, as it's the universal repository.

Linux Mint

Both Ubuntu and Mint use similar packaging management systems, making life much easier. Installation instruction for Ubuntu can also be used in Mint. A deadsnakes PPA working well with Mint.

Debian

Some evidence shows Ubuntu's 16.10 method can also work for Debian, however, there is no path for it to work on Debian 9. But you can make Python from the following source.

One disadvantage of Debian is, it doesn't install Sudo commands by default. Before installing Debian, you are required to check a compiled source from the Python instruction:

$-su.

$apt-get-install sudo

$-vi *etc*sudo

Then use sudo vim commands or any preferred text editor to open *etc*sudoers file. Adding the following text line at the end and changing "your_username" with the real usernames:

your_usernameALL=(ALL)ALL

Open SUSE

There are various websites which outline ways of getting zypper to add the newest Python; however, they might be outdated. They don't successfully work, and it's best to build Python from the source. Begin by installing the development tool that is completed in YaST through the menu or the zypper.

$sudu zypper install-tpatterndevel_C_C++

The procedure will take some time, and it involves installing 154 packages; however, when completed, you will be able to build the source.

Arch Linux

It's fairly strict to keep up with the Python release. Arch Linux might be preinstalled, and if it is, it is not the newest version, use the following command:

$packman -S Python

Running Python

Python can be run on a system in three ways.

Interactive Interpreter

You can start by entering Python and then begin programming in its interpreter by beginning from the command line on any platform that provides a command-line interpreter or a shell window.

A list of command-line options is given in the table below.

Option	Description
-d	Provide the output after debugging
-O	Optimized byte code generation *i.e.* the .pyo file is generated
-S	Don't run the import site for searching Python paths in a startup.
-v	Details of the import statement
-X	Disable the class-based built-in exceptions
-c cmd	It runs Python script sent in cmd String
File	The python script is run from the given file

Script from Command-line

Calling an interpreter on your application can help you run and execute your Python script in the command line.

Integrated Development Environment

It is possible to run Python from a GUI too. The one thing you require is a system that supports Python.

CHAPTER 2:

Which Version Is The Easiest To Use For Beginners?

Picking Your Version of Python

Before we dive too far into some of the steps that you need to take in order to work with the Python language and get it installed on your personal operating system, we first need to discuss some of the basics of the Python versions.

Python is relatively new, but it has still been out for a number of years, which means that there have been some updates to it, and there is more than one version that you are able to pick out.

This is good news for you because it not only gives you some options on what you would like to use, but it also shows us that Python is continually being developed, bringing in new features and things that you are sure to enjoy. It can just be a bit confusing when you are a beginner to figure out which version out of all the options is the best for your needs.

To keep it simple, there are two main versions of the Python language that you can work with. The older versions are the Python 2.X. There are a few versions of this that are still in use, and it can certainly bring you a lot of the power and more that you are looking for in a coding language. However, most programmers have moved on to working with Python 3.X because it is newer, it works better with some of the libraries that you may want to use with things like machine learning, and it has gotten rid of some of the big bugs and issues of the versions.

This doesn't mean that there isn't any value to working with the Python 2 versions that are out there, just be aware if you choose this that a lot of programmers have moved on to the newer versions, and a lot of the codes that you will see written out in Python are going to follow this new format. We are going to take a look at both versions and see what they are all about so you can make the decision that is right for your needs.

Python 2.X

The first option that we need to take a look at is the Python 2.X versions. This is one of the first versions of Python that came out, and it was released in 2000. Over the years, there have been a lot of improvements to this version of Python, and there are still programmers out there who want to work with this version of Python rather than some of the other options.

Python 2.7 is one of the newest versions of this kind, and it was released back in 2010. Even though there aren't plans for more releases of this kind of Python since most developers are working with Python 3 instead, it is still a good one to choose and there are some important reasons why this is the best one for us to choose to work with.

First, if you have any familiarity with 2.X or it is already installed on your computer, it is an easy version to work with and can save some time and hassle. It can do a lot of the coding that you want, and it has many of the features that you will need, without a lot of extras that could slow down the system.

You may also have a few programming needs in your organization that would do better with older technology, then the 2.X is the best option. For example, if your organization has policies that discourage or ban the installation of unapproved software from outside sources, then this version may be the best one for you to choose from. Many times, this version of Python was already installed on the system so this is the one that you would want to use.

In addition, there are many third-party libraries and packages that are used to help extend what capabilities the 2.X version can handle, and some of them are not present in the newer 3.X version. If you want to work with a specific library for your application, you may find that it is only available in the 2.X version. You would need to download this version to get it to work for you.

If you are looking through some of these reasons and you think that Python 2 is one of the best options for what you want to complete, it is still important to look through the Python 3 version and see what is there. There are going to be a few differences when it comes to best coding practices with these, and you want to make sure that you learn what these are before you start with any of the coding you would like to complete.

Python 3.X

Now that we have had a bit of time to take a look at the Python 2 version of this coding language, we also need to take a few minutes to look at Python 3. This is often the one that you are going to see when it is time to code in Python, and a lot of programmers have already moved over to using this version because it has a lot of the features and more that you are looking for when it comes to coding. Most of those who are getting started with learning how to use Python is also going to work with this version.

Python 3 in its earliest forms was released back in 2008, and there have been a number of improvements since that time. It is likely that we are going to see some more versions of this come out over the years because it is the version that most people are working on and developing right now.

Because this version is the most current language, most of the examples that we do with codes are going to be done with the Python 3 version. It has the latest libraries, the most third-party add-ons, and all the features and developments that you would need with your coding. However, both versions are easy to work with and you may find that you prefer one version over another for your programming needs.

Python 2.x vs. Python 3.x

There are two popular and official versions of Python: Python 3.x and 2.x. As of this writing, you can download Python 3.7.0 if you want the 3.x version. You can also download Python 2.7.15 if you want the 2.x version.

However, to prevent any conflicts and misunderstandings, please download and use Python 3.x. All the examples and lessons in this book are written with Python 3.x in mind.

The 2.x version is an older version of Python. Ever since the Python developers proceeded in developing Python 3.x, they have made a lot of changes to the behavior and even the syntax of the Python programming languages.

For example, if you divide 3 and 2 using the '/' operator in Python 2.x, you will receive an output of 1. If you divide the same numbers with the same operator in Python 3.x, you will receive an output of 1.5.

You may ask: If Python 3.x is new and improved, why are the developers keeping the old versions and why is Python 2.x being used?

The quick answer to that is *code migration*. Because there are many differences between the version 2.x and the version 3.x. Programs and scripts created using version 2.x need to be recoded to become compatible with version 3.x Python.

If you are dealing with a small program using version 2.x, then the code migration will be a trivial problem at best. However, if you have programs with thousands of lines, then migration can become a huge problem. Other issues with migrating to Python 3.x are code maintenance and retraining programmers to adapt to the changes.

Because of the aforementioned reasons, developers with huge programs written and ran using the version 2.x runtime environment did not bother making the transition to version 3.x.

CHAPTER 3:

Basics of Python Programming

What is Python Programming?

This is a programming language that is object-oriented and of high level and uses semantics. It is a high level in terms of structures in data and a combination of dynamic typing and binding. This is what makes it attractive to be used for Rapid Application Development and for connecting different elements.

Python with its simplicity and learning with ease helps in reading the programming language, and that is why it reduces the cost to maintain the program. Python encourages the program modularity and code reuse; this is because it supports different packages and modules. The standard library and the Python interpreter can be found in binary form. It is not necessary to charge all the available platforms and can be distributed freely.

Most programmers love the Python program because they offer great productivity. The edit-test debug is a cycle that is fast and does not need any compilation process. It is easier to debug a Python program; it will not cause any segmentation fault. An exception is raised when an error is discovered by the interpreter. When the exception is not known by the program, the interpreter prints a trace. The debugger, on a level of sourcing, will allow being inspecting any variables. There will be a settling of breakpoints, arbitrary expressions, and stepping on the code at any time. The Python is what writes the debugger, the easier, and a quick debugging method and programs of adding prints on the source and statements.

Python is open-source; this means you can use them freely for any commercial applications. Python is programmed to work on UNIX, Windows, and Mac and can be transferred to Java. Python is a language that helps in scripting and helps in web applications for different content.

It is like Perl and Ruby. Python is helped by several imaging programs; users are able to create customized extensions. There are different web applications supporting Python API like Blender and GIMP.

This information given on Python programming is beneficial for both the newbies and the experienced ones. Most of the experienced programmers can easily learn and use Python. There is an easier way to install Python. Most distributors of UNIX and Linux have the recent Python. That is the reason why most computers come already installed with Python. Before you start using Python, you need to know which IDEs and text editors' best work with Python. To get more help and information, you can peruse through introductory books and code samples.

The Python idea was discovered in 1980 after the ABC language. Python 2.0 was introduced, it had features like garbage collection and list comprehensions; which are used in reference cycle collection. When Python 3.0 was released in 2008, it brought about a complete language revision. Python is primarily used for developing software and webs, for mathematics and scripting systems. The latest version of Python is known a Python 3 while Python 2 is still popular. Python was developed to help in reading and similar aspects to different languages like English and emphasis on Mathematics. A new line is used to complete a Python command, as opposed to other programming languages that normally use semi-colons. It depends on indentation, whitespace, and defining the scope.

How to Use Python Programming

Before using Python, you first need to install and run it on your computer, and once you do that, you will be able to write your first program. Python is known as a programming platform that cuts across multiple platforms. You can use it on Linux, macOS, Windows, Java, and .NET machines freely and as an open-source. Most of the Linux and Mac machines come preinstalled even though on an outdated version. That is the main reason why you will need to install the latest and current version. An easier way to run Python is by using Thonny IDE; this is because it is bundled with the latest version of Python. This is an advantage since you will not need to install it separately.

To achieve all that, you can follow the simple steps below:

First, you will need to download Thonny IDE.

Then run the installer in order to install it on your computer.

Click on File option, then new. Save the file on .py extension, for instance, morning.py or file.py. You are allowed to use any name for the file, as long as it ends with .py. Write the Python code on the file before saving it.

To run the file, click on RUN, the run current script. Alternatively, click on F5 in order to run it.

There is also an alternative to install Python separately; it does not involve installing and running Python on the computer. You will need to follow the listed steps below:

Look for the latest version of Python and download it.

The next step is to run the installer file in order to install Python.

When installing, look for Add Python to environment variables. This will ensure that Python is added to the environment variables and that will enable you to run Python from any computer destination and part. You have the advantage to choose the path to install Python.

When you complete the process of installing, you can now run Python.

There is also an alternative and immediate mode to run Python. When Python is installed, you will type Python on the command line; the interpreter will be in immediate mode. You can type Python code, and when you press enter, you will get the output. For instance, when you type 1 + 1 and then press enter, you will get the output as 2. You can use it as a calculator, and you quit the process, type quit, then press enter.

The second way to do it is by running Python on the Integrated Development Environment. You can use any editing software in order to write the Python script file. All you need to do is to save it the extension .py, and it is considered a lot easier when you use an IDE. The IDE is a feature that has distinctive and useful features like file explorers, code hinting, and syntax checking and highlighting that a programmer can use for application development.

You need to remember that when you install Python, there is an IDE labeled IDLE that will also be installed.

That is what you will use to run Python on the computer, and it is considered the best IDE for beginners. You will have an interactive Shell when IDLE is opened. This is the point where you can have a new file and ensure that you save it as a .py extension.

Who Can Use Python Programming?

There is a big challenge out there in choosing a programming language that you can use for your coding businesses. The bigger question is which language are you supposed to learn?

Python is a program that is easy to use, and there are known companies that use it. This is one of the reasons why you should adapt to its uses. This is also the reason why worldwide developers have taken advantage.

Google

Since the beginning of Python, Google has been its supporter. They choose Python because it was easy to maintain it, deploy it, and faster in delivery. The first web-crawling spider used for Google was in Java 1.0. It was difficult to use and maintain, and they had to do it again on Python. Python is one of the main programming languages that Google uses, the others include Java, C++, and Go that are used for production. Python is an important part of Google. They have been using it for many years, and it remains a system that is evolving and grows. Many engineers that work for Google prefer using Python. They keep seeking engineers with Python skills.

Facebook

Production engineers that work for Facebook have a positive comment about Python. This has made Python among the top three programming languages after C++ and Hack. Facebook adopted Python because it is easy to use. With over 5000 services on Facebook, this is definitely the best programming language. The engineers do not need to maintain or write much coding, and this allows them to focus on live improvements.

This is one of the reasons why Facebook infrastructure scales efficiently. Python is used for infrastructure management, for network switch setup, and for imaging.

Instagram

From 2016, engineers working for Instagram declared that they were running the biggest Django web framework that was entirely written in Python. The engineers stated that they like Python because of the simple way to use it and because of how practical it is. That is why the engineers have invested their resources and time in using Python in all their trades. In recent times, Instagram has moved their codes from Python 2.7 to Python 3.

Spotify

Spotify is a music-streaming platform that uses Python as its programing language for back-end services and data analysis. The reason why Spotify decided to use Python is that they like the way it works in writing and in coding. Spotify will use its analytics in order to offer its users recommendations and suggestions. For the interpretation, Spotify uses Luigi that collaborates with Hadoop. The source will handle the libraries that work together. It will consolidate all the error logs and helps in troubleshooting and redeployment.

Quora

Before implementing their idea, Quora decided to use Python programming for their question and answer platform.

The Quora founders decided to go for Python because it was easy to read and write it. For great performance, they implemented C++. Python is still considered because of the frameworks it has like Pylons and Django.

Netflix

Netflix uses Python programming language to help in data analysis from their servers. They also use that in coding and other Python applications. It uses Python in the Central Alert Gateway and tracking any security history and changes.

Dropbox

This cloud storage system uses Python for the desktop client. Their programs are coded in Python. They use different libraries for Windows and Mac. And the reason being it is not preinstalled on any Mac or Windows and the Python version differs.

Reddit

Python programming language was used to implement Reddit. They choose Python because it has different versions of code libraries, and it was flexible to develop it.

What Can You Do With Python Programming?

There are numerous applications for Python programming, like machine learning, data science, and web development.

In addition, other several projects can use Python skills:

- With Python programming, you are able to automate boring stuff; this is the best approach for beginners. It helps with spreadsheet updates and renaming files. When you get to master Python basics, then this is the best point to start with. With the information, you will be able to create dictionaries, web scraping, creating objects, and working on files.

- Python will help you stay on top of the prices that are set on Bitcoin. Bitcoin and cryptocurrency have become a popular investment; this is because of its price fluctuation. In order to know the right move in regard to Bitcoin, you will need to be alert on their prices. With Python, it is possible to create a price notification for Bitcoin. This is the best way to start on crypto and Python.

- When your intention and plan is to create a calculator, the Python is the best programming language. You will be able to build back-end and front-end services, which are the best when it comes to deployment. It is important to create applications that users can easily use. If your interest is in UX and UI design, then Python has a graphical user interface that is easy to work with.

- Python is the best programming language to use when mining data from Twitter. With the influx of technology and the internet, it is easy to get data and information easily. Data analytics is a very important concept; it involves what people are talking about and their behavioral patterns. To get all the answers, Twitter is the best place to start with when your interest is in data analysis. There is a data-mining project on Twitter, and that is when your Python skills will come in handy.

- You will have the ability to create a Microblog with a flask. In recent times, most people have a blog. But again, it is not a bad idea to have your personal hub online. With Instagram and Twitter, microblogging has become a popular concept. With Python skills, you will be able to create your own microblog. When you are into web development, you do not need to be worried about knowing Flask. You can learn about it online and then move to Django which helps in web applications on a large scale.

- With Python skills, it is possible to build a Blockchain. The main reason for the development of blockchain was financial technology, even though it is spreading to other industries. As of now, blockchains can be used for any type of transaction, like medical records or real estate. When you get to build one for yourself, you will understand it better. You need to remember that, blockchain is not just for the individuals who are interested in crypto. When you build one, you will have a creative way for technology implementation to your own interest.

- You can bottle your Twitter feed with your Python skills, and this will help in web applications. You can create a simple web app that can help in navigation on your Twitter feed. You will not be using Flask, but rather Bottle; a low-dependency approach that is easy and quick to implement.

- There are PyGames that are easy to play with Python skills. You can use the skills to code several games and puzzles. With the Pygame library, it is easier to create your own games and developing it. It is an open and free source with computer graphics and sound libraries, and it helps in adding up interactive functions in the application. There are different games that can be used for library creation.

- With Python, it is possible to create something in relation to storytelling. Since the language is easy to use, that is why it creates a better environment for development and interaction.

Importance of Python Programming in the Economics

Most people ask if it is important for economists to learn any programming language. The answer is that it is important since they will use the skills to test and crush data sets. Most of the economist use Python as their main programming language to help with efficiency in order to run complex models. The idea of data analysis is used by other professionals not only by data scientists. Economists learn how to code in order to enjoy the ability to handle bigger data software. Large data can now be handled in spreadsheets when you use the new systems and all that can be done in a shorter time. That is the main reason why economists are adapting to the Python programming language.

Big data is what is used by different people all over. It helps in coding and for market and business intelligence. More spectrums of solutions are created by tech individuals, and they are not just for data scientists. This is why many economists are using and learning programming languages. Economists have been adapting to Python at a slower rate since they did not mostly depend on the data as compared to data scientists. They have adapted to the programming languages because of its flexibility, the breadth of functionalities, development, speed in computing, and the ability to operate between different systems.

Economists will deal with data that is on both low and high frequency. This is because of the increase in digitization and computing. The modern economy brings about greater power in computing and data sources. Years back, coding was used for only back office works, unlike recent times when it is used for front functions, and that is the reason why the use of Python has increased.

Importance of Python Programming at the Workplace

Several benefits come with learning Python if you have not learned the language. There is no need to panic because Python is a program that is easy to learn and can be used to learn other programming languages. You will understand the importance of Python since it is adopted by different companies like Instagram, Disney, Nokia, IBM,

Pinterest, and Google. When you learn about Python, you will have the skills needed to succeed and make good money at it.

Python programming language helps in developing prototypes, and the reason is that it is quick to learn and work with.

Most of the data mining and automation rely on Python. The reason being that language is better for general tasks.

With Python, you will get a better and productive environment for coding, unlike what most programming languages like Java and C++ will do. Most coders claim that with Python, they are better organized and productive when it comes to their work.

Since Python is not complicated and any beginner can easily read, learn, and understand, this means that anyone can work with the programming language. All that you will need is patience and practice to excel with the language. Python helps most programmers and development in large-scale dimensions.

Django is an open-source used in web development and application that is powered by Python. With Python, it is easier in improving the maintenance and readability of codes. Python helps in securing coding with updates and maintenance. The reason being it helps in developing quality in the software application. You will also be able to demonstrate all the concepts when using syntax rules. The quality that Python offers in terms of maintenance and readability makes it the best programming language. You can even use English keywords instead of punctuations.

Python helps with multiple paradigms that help a programmer know what is relevant in the work environment and requirement. Since Python supports different paradigms in programming, it is capable to feature different concepts that relate to functional programming. With Python, it is possible to develop a software application that is complicated and large.

Python help in integrating with different operating systems and interpreting different codes.

There is a possibility to redevelop this application without recompiling.

Python is able to provide better results as compared to the other programming language because of its library that is robust and big. From the library, it is possible to select the best as per the requirements and add more functionality to that. The feature will prevent having any additional code writing.

Python helps in simplifying any complicated software. This programming language helps in data analytics and visualization in any program that is developing.

When you are familiar with Python, you get to complete complex solutions without putting in a lot of time.

How Can You Earn Using Python Programming?

Blogging

Python programming helps in creating a blog, and the blog is used in making money. There are different types of blogging. You can specialize in programming as your niche in blogging. There are numerous ways to use your blog as a programmer. This includes online coding, charging your premium content, and affiliate links. Ensure all your content is SEO friendly with the relevant keywords that are what is used in ranking your page. In addition, SEO optimization should be on both on and off the page. You will be guaranteed of traffic to the site.

Apps Development

Your programming skills will be beneficial when you develop an app and monetize it. This programming helps in attaining that. Ensure that you market your app and use the automatic coding apps that will help in creating the app in a few minutes. With a great marketing technique, you will be able to make money out of app development. In order to sell it, you need to launch it on the App store.

Freelancing

This is a situation whereby you offer your services online. You should ensure that you look for the available online platforms and what works for you. You will be able to work from your own work schedule and make money while doing that.

Some of the genuine and known platforms include Guru, Freelancer, and Upwork.

You can also pitch directly to clients and offer your programming services.

You Can Earn Using Your Python Skills When You Make a Plugin

The other alternative is having a theme on WordPress. The best way to do it is by developing many apps and smaller modules like themes and WordPress plugins.

This is a great way to make money coding online. Most websites use WordPress, so if you can create WordPress plugins, you are assured of making a lot of money.

Another Way Is To Be an Online Educator and Start Selling Your Online Courses

All this is possible when you use your Python skills and do coding. Most people are adopting online courses where people who cannot attend classes can still learn.

When you have a personal website, you can offer free courses and tutorials and have traffic.

You can teach many students in economics and finance on how to code, and the Python skills.

When You Join Coding Competitions, You Will Be Using Your Python Skills and Still Earn Some Good Money

You can do development, data science, and design. When you are a winner, you are paid and get access to big companies who are on the lookout for competent coders.

Your Python Skills Will Help in Website Creation

You can share your programming tips and then display what you have. Ensure that your website shows all your skills, your bans, and your portfolio as a coder. When you have your brand established, you will get more clients that will be willing to get your services and consultation, and you can charge for that. A website can bring your earnings through Google Adsense, affiliate marketing, and sponsored ads.

The Terms Used in Python Programming

As it is the case for any skill, before going full-fledged into practicing the mode, it's important to learn the basic terms that are used in that domain.

To better understand your domain, you should learn the terms. For a beginner in Python Programming, we bring a few essential terms that you can be your learning 101 guides.

Below goes the programming terminology for beginners:

Algorithm

A set of rules that are created to solve an exact error. Tan error can be complex, such as converting video files to a different format or simply such as adding two numbers.

Program

This is an organized collection of instructions that performs a specific function when executed. It is processed by the CPU, an acronym for the central processing unit of the computer, before being executed. Microsoft Word is an example of a program that enables users to create and edit documents. Also, the browsers used are programs that are created to help users to browse the internet.

API

API is an acronym for Application Programming Interfaces. Sets of rules and procedures for building software applications. The APIs help with communication with third party programs, which is used to build different software. Major companies like Facebook and Twitter frequently use APIs to assist the developers to easily gain access to their services.

Bytecode

Python combines the source code into bytecode, an internal presentation of the Python program in a CPython interpreter.

Basically, the bytecode is an intermediate language running a virtual machine. The virtual machine is converted into machine code for it to execute it; however, the one-byte cannot run on a different virtual machine.

Bug

A bud is a term used to refer to an unexpected error in hardware or software, which causes it not to function. Bugs are often regarded as small computer glitches; however, bugs are life-threatening conditions and causing substantial financial losses. That's why it is important to focus on the process of finding bugs before programs in the applications, and this process is called testing.

Code

This is a term used to describe a written set of rules that are written using protocols of different languages like Python or Java. And also an informal use of the code describing text that is written in a specific language, and the reference code can be made for different languages such as CCS Code or HTML Code.

Command-Line Interface

This is a user interface that is based on the text, and it is used in viewing and managing computer files. The interface is also referred to as the command-line user, character user, and console users. In the early 60s, 70s, and 80s, the primary means of interaction with computers on terminals was the command-line interface.

Compilation

Creating an executable program by writing the code in a compiled programming language is known as compilation. With compiling, the computer understands the program and runs it without using the programming software that was used to create it. The compiler translates the computer programs that were written using numbers and letters to a machine language program. C++ is an example of a compiler.

Constants

This is also referred to as Const, which is a term that describes a value that doesn't change through the execution of the program as opposed to a variable. Constant is fixed and cannot be changed; it can be a string, number, or character.

Data Types

This is a group of a particular type of data. A computer cannot differentiate between a name or a number as a human, so it uses a special internal code to know the difference in the types of data it receives and how to process it.

There are various data types, which include character, which is the alphabets, the boolean values are the TRUE or FALSE, the integer is the numbers, and the floating-point number is the decimal numbers.

Array

The array is a list of a grouped type of data values, and the values have the same data type; however, they are different by the positioning in the array. For example, the age of students in a class is an array because they are all numbers, and also, the student's names in a class are array because of it is a character data type.

Declaration

This is a statement describing a variable, function, or other identifiers. It helps the compiler to identify the word, understand its meaning, and how to continue the process. They are essential, however, optional and are useful depending on the type of programming language.

Exception

The unexpected and special condition that is encountered during the execution of a program. This is also an error or a condition that changes the program to a different path. For example, when a program loads a file from the disk but the file does not exist. In order to avoid any fatal error, the highly important to handle and eradicate the exceptions in the program code.

Coroutine

A subroutine enters one point and exits in another point while a coroutine is a generalized meaning; it enters, exits, and also resumes at many different points. A coroutine is implemented with the async def statement.

Generic Function

Multiple functions are implementing a similar operation for different types. A dispatch algorithm will decide on which implementation to be used during a call.

Python Expression

A piece of code that is evaluated to a value. It's a collection of expression elements such as function calls, names, operators, and literals. An if-statement is not an assignment or an expression because it doesn't return a value.

Python Decorator

A function that is returning another function. It joins functionality without modifying it.

Loop

It is a series of instructions repeating a similar process that continues until a condition is completed, and it receives a command to stop. Then a question is asked on the program, and an answer will command the program to act, and then the loop continues to achieve a similar task. The process continues until there is no required action, and the code proceeds on. Loops are one of the most straightforward and powerful concepts in programming.

CHAPTER 4:

Python Data Types

A programming application needs to store a variety of data. Consider the scenario of a banking application that needs to store customer information. For instance, a person's name and mobile number; whether he is a defaulter or not; a collection of items that he/ she has loaned, and so on. To store such a variety of information, different data types are required. While you can create custom data types in the form of classes, Python provides six standard data types out of the box. They are:

- Strings

- Numbers

- Booleans

- Lists

- Tuples

- Dictionaries

Python treats a string as a sequence of characters. To create strings in Python, you can use single as well as double-quotes. Take a look at the following script: first_name = 'mike' # String with single quotation last_name = " johns" # String with double quotation full_name = first_name + last_name # string concatenation using + print(full_name) In the above script we created three string variables: first_name, last_name and full_name. A string with single quotes is used to initialize the variable "first_name" while a string with double quotes initializes the variable "last_name." The variable full_name contains the concatenation of the first_name and last_name variables. Running the above script returns the following output: mike johns

Numbers

There are four types of numeric data in python:

int (Stores integer e.g 10)

float (Stores floating point numbers e.g 2.5)

long (Stores long integer such as 48646684333)

complex (Complex number such as 7j + 4847k)

To create a numeric Python variable, simply assign a number to a variable. In the following script, we create four different types of numeric objects and print them on the console. int_num = 10 # integer float_num = 156.2 #float long_num = -0.5977485613454646 #long complex_num = -. 785 + 7J #Complex print(int_num)

print(float_num)

print(long_num)

print(complex_num)

The output of the above script will be as follows:

Boolean

Boolean variables are used to store Boolean values. True and False are the two Boolean values in Python. Take a look at the following example:

defaulter = True

has_car = False

print(defaulter and has_car)

In the script above we created two Boolean variables "defaulter" and "has_car" with values True and False respectively. We then print the result of the AND operation on both of these variables. Since the AND operation between True and False returns false, you will see false in the output.

Lists

In Python, the List data type is used to store the collection of values. Lists are similar to arrays in any other programming language. However, Python lists can store values of different types. To create a list opening and closing square brackets are used. Each item in the list is separated from the other with a comma. Take a look at the following example.

cars = [' Honda', 'Toyota', 'Audi', 'Ford', 'Suzuki', 'Mercedez']
print(len(cars)) #finds total items in string print(cars)

In the script above we created a list of named cars. The list contains six-string values *i.e.* car names. Next, we printed the size of the list using the len function. Finally, we print the list on the console.

Tuples

Tuples are similar to lists with two major differences. Firstly, opening and closing braces are used to create tuples instead of lists that use square brackets. Secondly, tuple once created is immutable which means that you cannot change tuple values once it is created. The following example clarifies this concept. cars = [' Honda', 'Toyota', 'Audi', 'Ford', 'Suzuki', 'Mercedez']

cars2 = (' Honda', 'Toyota', 'Audi', 'Ford', 'Suzuki', 'Mercedez')

cars [3] = 'WV'

cars2 [3] = 'WV'

In the above script, we created a list named cars and a tuple named cars2. Both the list and tuple contains a list of car names. We then try to update the third index of the list as well as the tuple with a new value. The list will be updated but an error will be thrown while trying to update the tuple's third index.

This is due to the fact that tuple, once created cannot be modified with new values.

Dictionaries

Dictionaries store the collection of data in the form of key-value pairs. Each key-value pair is separated from the other via comma. Keys and values are separated from each other via the colon. Dictionary items can be accessed via index as well as keys. To create dictionaries, you need to add key-value pairs inside the opening and closing parenthesis. Take a look at the following example.

cars = {' Name':' Audi', 'Model': 2008, 'Color':' Black'}

print(cars[' Color'])

print(cars.keys())

print(cars.values())

In the above script, we created a dictionary named cars. The dictionary contains three key-value pairs *i.e.* 3 items. To access value, we can pass the key to the brackets that follow the dictionary name. Similarly, we can use keys() and values() methods to retrieve all the keys and values from a dictionary, respectively.

Decimal Mode

This is a Python standard module library. Before using it, you need to import this module with import instructions before using it. After correctly importing this module, we can use the decimal. Decimal class to store accurate numbers. If the parameter is not an integer, we must pass in the parameter as a string.

For example:

Import Decimal

Num = decimal.decimal (" 0.1") + decimal.decimal (" 0.2")

And the result will be 0.3. Use the round () function to force the specified number of decimal places round(x[, n]) to be a built-in function, which returns the value closest to parameter x, and n is used to specify the number of decimal places returned.

For example:

Result = 0.1 + 0.2. The program statement above print (round(num, 1)) takes the variable num to one decimal place, thus obtaining a result of 0.3. 2.2.3. Boolean Data Type (bool) is a data type that represents the logic and is a subclass of int, with only True value (true) and False value (false). Boolean data types are commonly used in program flow control. We can also use the value "1" or "0" to represent true or false values. For example, the string and integer cannot be directly added, and the string must be converted to an integer. If all the operations are of numeric type, Python will automatically perform type conversion without specifying the forced conversion type.

For example:

num = 5 + 0.3

Result num = 5.3 (floating-point number)

Python will automatically convert an integer to floating-point number for operation. In addition, Boolean values can also be calculated as numeric values. True means 1, False means 0.

For example:

num = 5 + True

result num = 6 (integer).

If you want to convert strings to Boolean values, you can convert them by the bool function. Use the print () function in the following sample program to display Boolean values.

[sample procedure: bool.py] converts bool type

print(bool(0))

print(bool(""))

 print(bool(" "))

print(bool(1))

The execution results of the 05

print(bool(" ABC ") sample

The program is shown. Program Code Resolution: Line 02: An empty string was passed in, so False was returned. Line 03 returns True because a string containing a space is passed in.

When using Boolean, it values False and True. Pay special attention to the capitalization of the first letter.

Constant

Constant refers to the value that the program cannot be changed during the whole execution process. For example, integer constants: 45, -36, 10005, 0, etc., or floating-point constants: 0.56, -0.003, 1.234E2, *etc.* Constants have fixed data types and values. The biggest difference between variable and constant is that the content of the variable changes with the execution of the program, while the constant is fixed. Python's constant refers to the literal constant, which is the literal meaning of the constant. For example, 12 represents the integer 12. The literal constant is the value written directly into the Python program. If literal constants are distinguished by data type, there will be different classifications, for example, 1234, 65, 963, and 0 are integer literal constants. The decimal value is the literal constant of the floating-point types, such as 3.14, 0.8467, and 744.084. As for the characters enclosed by single quotation marks (') or double quotation marks ("), they are all string literal constants. For example," Hello World " and " 0932545212 " are all string literal constants.

Formatting Input and Output Function

In the early stage of learning Python, the program execution results are usually output from the control panel, or the data input by the user is obtained from the console. Before, we often use the print () function to output the program's execution results. This section will look at how to call the print () function for print format and how to call the input () function to input data.

The Print Format

The print () function supports the print format. There are two formatting methods that can be used, one is a print format in the form of "%" and the other is a print format in the form function. "%" print format formatted text can use "% s" to represent a string, "% d" to represent an integer, and "% f" to represent a floating-point number.

The syntax is as follows:

PRINT (formatted text (parameter 1, parameter 2, ..., parameter n))
For example:

score = 66 Print (" History Score: %d"% score")

Output

Result: History Score: 66 %d is formatted, representing the output integer format. The print format can be used to control the position of the printout so that the output data can be arranged in order.

For example:

print("% 5s history result: %5.2f"% (" Ram," 95)). The output results of the sample program print("% 5s history results: %5.2f"% (" Raj," 80.2)).

The formatted text in the above example has two parameters, so the parameters must be enclosed in brackets, where %5s indicates the position of 5 characters when outputting, and when the actual output is less than 5 characters, a space character will be added to the left of the string. %5.2f represents a floating-point number with 5 digits output, and the decimal point occupies 2 digits.

The following example program outputs the number 100 in the floating-point number, octal number, hexadecimal number, and binary number format using the print function, respectively.

You can practice with this example program:

[Example Procedure:

print_%. py]

Integer Output

visual = 100 in Different Decimal Numbers

print (" floating point number of number %s: %5.1f"% (visual, visual))

\print (" octal of number %s: %o"% (visual, visual)) print (" hex of number %s: %x"% (visual, visual))

The execution result of the print (" binary of number %s: %s"% (visual, bin(visual))) will be displayed.

Program Code Analysis

Lines 02-04: output in the format of floating-point number octal number and hexadecimal number.

Line 05: Since binary numbers do not have formatting symbols, decimal numbers can be converted into binary characters through the built-in function bin () and then output.

The Output

Print format of the format () function can also be matched with the format () function. Compared with the% formatting method, the format () function is more flexible. Its usage is as follows: print("{} is a hard-working student ... ." format (" First ranker ")). Generally, the simple FORMAT usage will be replaced by the braces "{}," which means that the parameters in FORMAT () are used within {}. The format () function is quite flexible and has two major advantages: regardless of the parameter data type, it is always indicated by {}. Multiple parameters can be used, the same parameter can be output multiple times, and the positions can be different.

For example: print("{ 0} this year is {1} years old ... ." format (" First ranker ," 18)), where {0} means to use the first parameter, {1} means to use the second parameter, and so on. If the number inside {} is omitted, it will be filled in sequentially.

We can also use the parameter name to replace the corresponding parameter,

For example: print("{ name} this year {age}.." format(name =" First ranker ," age = 18)) can specify the output format of the parameter by adding a colon ":" directly after the number. For example: print('{ 0:. 2f}'. format(5.5625)) means the first parameter takes 2 decimal places. In addition, the string can be centered, left-aligned, or right-aligned with the "<" ">" symbol plus the field width.

For example:

print("{ 0: 10} score: {1: _ 10}." format (" Ram," 95)) print("{ 0: 10} results: {1: > 10}." format(" Raj," 87))

The output of the print("{ 0: 10} result: {1:* < 10}." format(" Ram," 100)) program is shown. {1: _ 10} indicates that the output field width is 10, and the following line "_" is filled and centered. {1: > 10} indicates that the output field is 10 wide and aligned to the right, and the unspecified padding characters will be filled with spaces. {1:* < 10} indicates that the output field is 10 wide, filled with an asterisk "*" and aligned to the left.

Input Function:

Input() input is a common input instruction, which allows users to input data from a 'standard input device' (usually refers to keyboard) and transfer the numerical value, character, or string entered by users to the specified variable.

For example, if you calculate the total score of history and mathematics for each student, you can use the input command to let the user input the results of Chinese and mathematics, and then calculate the total score.

The syntax is as follows:

variable = input (prompt string) when data is Entered, and the enter key is pressed, the entered data will be assigned to the variable. The "prompt string" in the above syntax is a prompt message informing the user to enter, for example, the user is expected to enter height, and the program then outputs the value of height.

The program code is as follows:

height = input (" Give exact your height:")

For example, score = input (" Give exact your math score:")

The output of the print("% s' math score: %5.2f"% (" ram," float(score))) When the program is executed, it will wait for the user to input data first when it encounters the input instruction. After the user completes the input and presses the Enter key, it will store the data input by the user into the variable score. The data input by the user is in a string format. We can convert the input string into an integer, floating-point number, and a bool type through built-in functions such as int (), float (), bool (). The format specified in the example is floating point number (% 5.2f), so call the float () function to convert the input score value into a floating point number. The next section will introduce a more complete data type conversion.

If we use an integrated development environment such as Spyder, don't forget to switch the input cursor to Python console before inputting when the program is executed to input prompt information.

Let's practice the use of input and output again through the sample program.

[Example Procedure: Format. Py] format.py】

name = input (" Give exact Name:")

che_grade = input (" Give a language score:")

math_grade = input (" Give Math Score:")

print("{ 0: 10} {1: > 6} {2: > 5}."

format (" name," "language," "mathematics"))

The execution results of the 06print ("{ 0: < 10} {1: > 5} {2: > 7}." format (name, che *grade, math* grade)).

Program Code Analysis:

Lines 01-03: Require users to enter their names, Chinese scores, and math scores in sequence. Lines 05 and 06: Output the names, Chinese and math headers in sequence, and then output the names and results of the two subjects in the next line.

Data type conversion requires operations between different types in expressions. We can convert data types "temporarily," that is, data types must be forced to be converted.

There are three built-in functions in Python that cast data types.

int ():

Cast to integer data type. For example: x = "5" num = 5 + int(x) Print(dude) # Result: The value of 10 variable x is "5" and is of string type, so int(x) is called first to convert to integer type.

float ():

Cast to floating point data type.

For example: x = "5.3"

dude = 5 + float(x)

Print(dude) # Result: The value of 10.3 variable X is "5.3" and is of string type, so float(x) is first used to convert to floating point type.

str ():

Cast to String Data Type

For example: first = "5.3"

dude = 5 + float(first)

Print (" The output value is" + str(dude)) # Result:

The output value is 10.3.

In the above program statement, the string of words:

"the output value is" in the print () function is a string type, the "+" sign can add two strings, and the variable dude is a floating point type, so the str () function must be called first to convert it into a string. [sample procedure: conversion.py]

Data type conversion

str = "{ 1} +{ 0} = {2}"

first = 150

second = "60"

The execution result of 04 print(str.format(first, second, first + int(second))) program

Line 01:

Since B is a string, specify its display format first.

Note that the numerical numbering sequence of braces' {}' is {1}, {0}, {2}, so the display sequence of variables A and B is different from the parameter sequence in Format. Line 04: First, call int () to convert b to an integer type, and then calculate.

Practice Exercise

The pocket money bookkeeping butler designed a Python program that can input the pocket money spent seven days a week and output the pocket money spent every day. The sample program illustrates that this program requires the user name to be entered, and then the sum of spending for each day of the week can be entered continuously, and the pocket money spent for each day can be output. The program code shows that the following is the complete program code of this example program. [example program: money.py] pocket money bookkeeping assistant # -*- coding: utf-8 -*- """ You can enter pocket money spent 7 days a week and output the pocket money spent every day. """ name = value (" Give name:") working1 = value (" Give the total amount of pocket money for the first working:")

working2 = value (" Give the total amount of pocket money for the next working:")

working3 = value (" Give the total amount of pocket money for the third working:")

working4 = value (" Give the total amount of pocket money for the fourth working:")

working5 = value (" Give the total spending of pocket money for the fifth working:")

working6 = value (" Give the total spending of pocket money for the sixth working:")

working7 = input (" Give the total allowance for the seventh working:")

print("{ 0: < 8}{ 1: ^ 5}{ 2: ^ 5}{ 3: ^ 5}{ 4: ^ 5}{ 5: ^ 5}{ 6: ^ 5}{ 7: ^ 5}." \

format(" name,"" working1,"" working2,"" working3," \ "working4,"" working5,"" working6," \ "working7")) print("{ 0: < 8}{ 1: ^ 5}{ 2: ^ 5}{ 3: ^ 5}{ 4: ^ 5}{ 5: ^ 5}{ 6: ^ 5}{ 7: ^ 5}." \ format(name, working1, working2, working3, working4, working5, working6, working7))

CHAPTER 5:

The Python Variables

The Python Variables

The Python variables are an important thing to work with as well. A variable, in simple terms, is often just going to be a box that we can use to hold onto the values and other things that show up in our code. They will reserve a little bit of the memory of our code so that we are able to utilize it later. These are important because they allow us to pull out the values that we would like to use at a later time without issues along the way.

These variables are going to be a good topic to discuss because they are going to be stored inside the memory of our code. And you will then be able to assign a value over to them and pull them out in the code that you would like to use. These values are going to be stored in some part of the memory of your code and will be ready to use when you need. Depending on the type of data that you will work with, the variable is going to be the part that can tell your compiler the right place to save that information to pull it out easier.

With this in mind, the first thing that we need to take a look at is how to assign a value over to the variable. To get the variable to behave in the manner that you would like, you need to make sure that a minimum of one value is assigned to it. Otherwise, you just save an empty spot in the memory. If the variable is assigned properly to some value, and sometimes more than one value based on the code you are using, then it is going to behave in the proper manner and when you call up that variable, the right value will show up.

As you go through and work with some of the variables you have, you may find that there are three main options that are able to use. Each of these can be useful and it is often going to depend on what kind of code you would like to create on the value that you want to put with a particular variable. The three main types of variable that you are able to choose from here will include:

Float

This would include numbers like 3.14 and so on.

String

This is going to be like a statement where you could write out something like "Thank you for visiting my page!" or another similar phrase.

Whole Number

This would be any of the other numbers that you would use that do not have a decimal point.

When you are working with variables in your code, you need to remember that you don't need to take the time to make a declaration to save up this spot in the memory. This is automatically going to happen once you assign a value over to the variable using the equal sign. If you want to check that this is going to happen, just look to see that you added that equal sign in, and everything is going to work.

Assigning a value over to your variable is pretty easy. Some examples of how you can do this in your code would include the following:

x = 12 #this is an example of an integer assignment

pi = 3.14 #this is an example of a floating-point assignment

customer name = John Doe #this is an example of a string assignment

There is another option that we are able to work with on this one, and one that we have brought up a few times within this section already. This is where we will assign more than one value to one for our variables. There are a few cases where we are going to write out our code and then we need to make sure that there are two or more values that go with the exact same variable.

To make this happen, you just need to use the same kind of procedure that we were talking about before. Of course, we need to make sure that each part is attached to the variable with an equal sign. This helps the compiler know ahead of time that these values are all going to be associated with the same variable. So, you would write out something like a = b = c = 1 to show the compiler that all of the variables are going to equal one. Or you could do something like 1 = b = 2 in order to show that there are, in this case, two values that go with one variable.

The thing that you will want to remember when you are working with these variables is that you have to assign a value in order to make the work happen in the code. These variables are also just going to be spotted in your code that is going to reserve some memory for the values of your choice.

CHAPTER 6:

Basic Syntax

Syntax refers to the set of rules that define the correct form and sequence of words and symbols in a programming language. This chapter will discuss the syntax and conventions used in Python3.

Python Keywords

Python has 33 keywords that are reserved for built-in functions and processes. You have to be aware of these keywords and avoid using them as an identifier in your programs.

True	False	None
while	for	return
continue	break	pass
if	elif	else
class	def	del
import	From	lambda
and	or	not

finally	nonlocal	global
with	Raise	as
assert	Except	yield
try	is	in

Naming Conventions for Identifiers

An identifier is a name given to a variable, string, dictionary, list, class, module, function, and all other objects in Python. Each type of object has its own naming convention.

In general, an identifier may consist of one letter or a combination of uppercase and lowercase letters, underscores, or the digits 0 to 9.

Examples of Identifiers:

X

x

my_dict

my_list9

UPPERCASE

lowercase

UPPERCASE_WITH_UNDERSCORES

lowercase_with_underscores

CamelCase or CapWords

mixedCase

You should not use keywords that are identifiers but if you really need to use one, you can distinguish it from a keyword by using a trailing underscore.

This is preferred over abbreviation or word corruption.

For example: global_, class_

Identifiers should not contain symbols or special characters such as #, $, %, and @.

Identifiers may end in a number but should never start with a number. For example, var_1 is valid but 1var is not.

A multiple-word identifier is acceptable but separating each unit with an underscore will make it more readable.

For example, my_main_function is more preferred over mymainfunction.

You can use the following style guide when naming specific identifiers:

Global Variables

Names of global variables are written in lower case. If there are two or more words in an identifier, they are preferably separated by an underscore for clarity and readability.

Classes

The UpperCaseCamelCase convention is used when naming a class. That is, the identifier should start with an uppercase letter. If there are two or more words in a class identifier, they all start in a capital letter and are joined together. For example: Employees, MyCamelCase, TheFamousList, Cars, MyCar

Instance Variables

Names of instance variables consisting of two or more words should be separated by underscore(s). They should be written in lowercase.

Examples: student_one, member_status, employee,

Functions

A function name should be written in lowercase. Multiple-word function names should be separated by an underscore for better readability.

Examples: multiplier, adder, sub_function

Arguments

Arguments must consist of letters in lowercase. The word 'self' is always the first argument in instance methods while 'cls' is the first argument in class methods.

Modules and Packages

Module identifiers should be in lowercase and are usually written as a short single-word name. Using multiple words is discouraged but you may do so to facilitate clarity and readability. A single underscore should separate each unit in a module and package identifier consisting of two or more words.

Constants

Constants are written in uppercase letters and separated by an underscore if the name consists of multiple words.

Examples: TOTAL, MAX_OVERFLOW

Quotation Marks

Quotation marks are used to indicate strings. You may use single ('), double ("), or triple (''') quotes but you must use the same mark to start and end string literals.

Examples:

'status', "Johnny", '''age limit: ''', 'Enter a vowel: '

Indentation

Python uses indentation or white spaces to structure programs or blocks of codes. You will not use curly braces {} to distinguish a group of related statements. Instead, you will use equal indentation to mark a block of code. For this reason, Python programs look neat, clutter-free, organized, and readable. A nested block is created by indenting a block further to the right.

While you can use tabs to indent a code, Python programmers commonly use four white spaces to mark a block of code. This is a language convention that you should consider for uniformity. It is not strictly imposed, however, and you may use tabs instead. Just don't forget to make sure that all lines in a code block are indented consistently if you want your program to run as expected. When using the built-in text editor, you will notice that it intuitively provides four white spaces whenever it is expected based on the most recently typed statement.

Here is a snippet of a program that will give you an idea of how Python programs are structured:

```
def Vehicle_Rental(days):

rate = 15 * days

if days >= 15:

rate -= 20
```

elif days >= 7:

rate -= 5

return fees

Statements

Statements are expressions that may be entered on the Python prompt or written inside a program and are read and executed by the interpreter. Python supports statements such as 'if, 'for', 'while', 'break', and assignment statements.

Multiple-Line Statements

Some statements may span over one line of code and stretch to several lines.

An implicit way to tell Python that the lines are related and form a single statement is by enclosing the statements in parentheses (), curly braces {}, or square brackets [].

For example, the following assignment statement creates a list and will require 3 lines of codes. The items are enclosed in square brackets which indicate that they belong to a list and are part of a single statement.

>>> my_collection = ['stamps', 'stationery', 'pens', 'compact discs', 'boxes', 'watches', 'metallic toy cars', 'numismatic coins', 'paintings', 'newspaper clippings', 'books']

You may also indicate continuation explicitly by using a backslash \ at the end of each line of related statements. For example: my_alphabet = 'a', 'b', 'c', 'd','e', 'f', 'g', 'h', 'i' \

'j', 'k', 'l' 'm', 'n', 'o', 'p', 'q', 'r' \

's', 't', 'u','v', 'w', 'x', 'y', 'z'

Documentation String

A docstring, short for documentation string, is used to provide information on what a class or a function does.

Docstrings are typically found at the top of a code block that defines a class, function, method, or module.

By convention, a docstring starts with a one-line imperative phrase that begins in a capital letter and ends with a period. It may stretch over multiple lines and are enclosed in triple-double quotes (""").

Here are examples of a docstring:

def adder(x, y):

   """Function to get the sum of two numbers."""

   return x + y

class Doc(obj):

   """ Explain function docstring.

A function name is introduced by the keyword def. Function definition statements end in a colon and are followed immediately by a docstring in triple double-quotes. The function body is distinguished by the white spaces right after the definition statement and docstring.

"""

Comments

Comments are notes written inside programs. They are intended to provide documentation about the programming steps, processes, and other information that a programmer may consider important. Comments are useful when evaluating a program. They facilitate a smooth review and transition between programmers. A hash # symbol is used to introduce a comment. It tells Python to ignore the line and proceed to the execution of the next statement.

For example:

#print a birthday greeting

print('Happy Birthday, Member!')

Comments may stretch over multiple lines. You may use a hash # symbol at the start of each line to wrap the lines together.

Example:

#This step is important

\# because it is the basis

\# of succeeding processes.

Another way to wrap long multiple-line comments is by using triple quotes at the beginning and end of the comment.

For example:

"""This is another way

to write comments that

span several lines."""

CHAPTER 7:

Classes in Python

Objects are often based on real-life objects, or they can be special objects created for your program. It all depends on what problem you are solving. First, let's look at the basic syntax for a class.

class ClassName:

'optional class documentation string'

class_suite

Class_suite represents the methods and variables in the class. The class documentation string can tell you what the class is.

Let's look at an example class. This class represents a person.

class Person:

'This class represents a person.'

name = "John Doe"

age = 20

def displayName():

```
print "My name is " + self.name

def displayAge():

print "My name is " + self.age

def changeName(self, newName):

self.name = newName

def changeAge(self, changedAge):

self.age = changedAge
```

This class represents a person and has a name and age. The functions, otherwise known as methods, in the class initialize the class, print the name, print the age, change the name, and change the age. These are often known as "getters" and "setters," which change the attributes of the object. The attributes, in this case, are the name and age of the person.

The first method __init__() is a special method. This is also known as the class constructor, which can be used to create a new object and pass in values for the attributes. In this case, you pass in values for the name and age that you want the person to have. If there aren't any values, then the defaults "John Doe" and age 20 are used.

The other class methods you want to use can be declared just like regular functions. The only difference is that the first argument is always self, which represents the current instance of the object. There is no need to include it when you call the methods, though. Python adds that for you. Let's look at an example of creating an object and calling these methods.

person1 = Person("John Smith", 30)

This creates a new person named John Smith who is 30-years-old.

person1.displayName()

person1.displayAge()

This will print the name and the age.

person1.changeName("Jane Smith")

person1.changeAge(31)

This will change the person's name to Jane Smith and make the age 31.

You can also add, modify, and delete attributes to an object at any time.

Let's add an occupation attribute to the person above.

person1.occupation = "Farmer"

Now, we have made person1 a farmer.

person1.occupation = "Retired"

Now, we have changed the person's occupation to "Retired."

You can use the following functions to modify attributes as well.

getattr(object_name, attribute_name)

This will return the value of attribute_name for object_name.

hasattr(object_name, attribute_name).

This will return true if object_name has attribute_name and false otherwise.

setattr(object_name, attribute_name, attribute_value)

This will set the attribute_name attribute to attribute_value for object_name. If the attribute doesn't exist, it will be created for you.

delattr(object_name, attribute_name)

This will delete the attribute_name attribute for object_name.

Let's look at an example with our person 1 object.

getattr(person1, name) will return Jane Smith because that is the current value for the name.

hasattr(person1, age) will return true because the person has an age.

hasattr(person1,degree) will return false because the person doesn't have a degree attribute.

setattr(person1, degree, "Bachelors") will create the attribute degree and set the value to "Bachelor's."

setattr(person1, age, 35) will change Jane Smith's age to 35.

delattr(person1, degree) will delete Jane Smith's degre atrribute.

Python also has some built-in attributes for classes.

__dict__ has a dictionary containing the class's namespace.

__doc__ has the class documentation string if there is one.

__name__ has the class name.

__module__ has the module name if there is one.

__bases__ this has the base classes that the class inherits from.

You can also delete an object with the del statement.

So, to delete person1, you can say:

del person1

Classes can inherit attributes, methods, and other behaviors from classes you have already made. There is no need to reinvent the wheel.

You can have one class inherit from multiple classes. This syntax looks like:

class SubClassName(ParentClass1):

'documentation string'

class_contents

This subclass, SubClassName, inherits from ParentClass1.

If you want to have multiple parent classes, that might look like this.

class SubClass2(ParentClass2, ParentClass3):

'documentation string'

class_contents

This SubClass2 inherits from ParentClass2 and ParentClass3.

Let's look at a Student class that inherits from a person.

class Student(Person):

major = "Undecided"

minor = "Undecided"

def __init__(self, major, minor):

self.major = major

self.minor = minor

def setMajor(self, newMajor):

self.major = newMajor

def setMinor(self, newMinor):

self.minor = newMInor

def printMinor():

print self.minor

def printMajor():

print self.major

There are some special functions that you can use when dealing with the subclasses.

issubclass(subclass, parentclass) will return true if the subclass inherits from the parent class and false otherwise.

isinstance(object, class) will return true if the object is in fact an instance of class and false otherwise.

(Tutorials Point)

Projects

Exercise 1: Write a class called Circle, which contains a radius. Create a set method and a print method for the radius.

Exercise 2: Create an instance of the circle class, add a pi attribute, and set it to 3.14.

Exercise 3: Create an employee class that inherits from a person. Give it a salary and occupation. Create a set method and print method for the salary and occupation.

Exercise 4: Create a counter class with a set method and a print method. Add methods to increment and decrement the counter by 1. Add methods to increment and decrement by 5.

Exercise 5: Create a floating-point number that can decrement and increment by 0.1 and 0.5.

Exercise 6: Create a double number class that can increment and decrement by 0.01 and 0.05. Also, add a print and set method for the counter.

Exercise 7: Create a price class. Increment and decrement by ten cents, one cent, one dollar, and five dollars.

Exercise 8: Create an animal class with a bark and fly function.

Answers

Exercise 1.

class Circle:

'This class represents a circle'

radius = 1

def __init__ (self, radius):

self.radius = radius

def setRadius(self, newRadius):

self.radius = newRadius

def printRadius(self):

print self.radius.

Exercise 2.

circle1 = Circle(5)

circle1.pi = 3.14

Exercise 3.

class Employee(Person):

salary = 0

occupation = "Unemployed"

def __init__(self, salary, occupation):

self.salary = salary

self.occupation = occupation

def setSalary(self, newSalary):

self.salary = newSalary

def setOccupation(self, newOccupation):

self.occupation = newOccupation

```
def printSalary(self):

    print self.salary

def printOccupation(self):

    print self.occupation
```

Exercise 4.

```
class Counter:

    'This class represents an integer counter.'

    counter = 0

    def __init__(self, count):

        self.counter = count

    def setCounter(self, newCount):

        self.counter = newCount

    def printCount(self):

        print self.counter

    def incrementByOne(self):

        self.counter = counter + 1

    def decrementByOne(self):
```

self.counter = counter - 1

def incrementByFive(self):

self.counter = counter + 5

def decrementByFive(self)

self.counter = counter - 5

Exercise 5.

class FloatCounter:

counter = 0.0

def __init__(self, count):

self.counter = count

def incrementByTenth(self):

self.counter = counter + 0.1

def decrementByTenth(self):

self.counter = counter - 0.1

def incrementByHalf(self):

self.counter = counter + 0.5

def decrementByHalf(self):

self.counter = counter - 0.5

def setCounter(self, newCount):

self.counter = newCount

def printCounter(self):

print counter

Exercise 6.

class DoubleCounter:

counter = 0.0

def __init__(self, counter):

self.counter = counter

def setCounter(self,newCounter):

self.counter = newCounter

def printCounter(self):

print counter

def incrementBy0.01(self):

counter = counter + 0.01

def decrementBy0.01(self):

counter = counter - 0.01

def incrementBy0.05(self):

counter = counter + 0.05

def decrementBy0.05(self):

counter = counter - 0.05

Exercise 7.

class Price:

price = 0.0

def __init__(self, price):

self.price = price

def printPrice(self):

print price

def setPrice(self, newPrice):

self.price = newPrice

def incrementByOneCent(self):

self.price = price + 0.01

def decrementByOneCent(self):

self.price = price - 0.01

def incrementByTenCents(self):

self.price = price + 0.1

def decrementByTenCents(self):

self.price = price - 0.1

def incrementByOneDollar(self):

self.price = price + 1.0

def decrementByOneDollar(self):

self.price = price - 1.0

def incrementByFiveDollars(self):

self.price = price + 5.0

def decrementByFiveDollars(self):

self.price = price - 5.0

Exercise 8.

class Animal:

type = "Animal"

def __init__(self, type):

```
self.type = type

def setType(self, newType):

self.type = newType

def bark(self):

print "Bark"

def fly(self):

print "Fly"
```

CHAPTER 8:

Conditions and Loops

Computing numbers and processing text are two basic functionalities that a computer program instructs a computer. An advanced or complex computer program has the capability to change its program flow. That is usually done by allowing it to make choices and decisions through conditional statements.

Condition statements are one of a few elements that control and direct your program's flow. Other common elements that can affect program flow are functions and loops.

A program with a neat and efficient program flow is like a create-your-own-adventure book. The progressions, outcomes, or results of your program depend on your user input and runtime environment.

For example, say that your computer program involves talking about cigarette consumption and vaping. You would not want minors to access the program to prevent any legal issues.

A simple way to prevent a minor from accessing your program is to ask the user his age. This information is then passed on to a common functionality within your program that decides if the age of the user is acceptable or not.

Programs and websites usually do this by asking for the user's birthday. That being said, the below example will only process the input age of the user for simplicity's sake.

```
>>> userAge = 12
>>> if (userAge < 18):
    print("You are not allowed to access this program.")
else:
    print("You can access this program.")
You are not allowed to access this program.
>>> _
```

Here is the same code with the user's age set above 18.

```
>>> userAge = 19
>>> if (userAge < 18):
    print("You are not allowed to access this program.")
else:
    print("You can access this program.")
```

You can access this program.

```
>>> _
```

The **if** and **else** operators are used to create condition statements. Condition statements have three parts. The conditional keyword, the Boolean value from a literal, variable, or expression, and the statements to execute.

In the above example, the keywords **if** and **else** were used to control the program's flow. The program checks if the variable **userAge** contains a value of less than 18. If it does, a warning message is displayed. Otherwise, the program will display a welcome message.

The example used the comparison operator less than (<). It basically checks the values on either side of the operator symbol. If the value of the operand on the left side of the operator symbol was less than that on the right side, it will return True. Otherwise, if the value of the operand on the left side of the operator symbol was equal or greater than the value on the right side, it will return False.

"if" Statements

The "if" keyword needs a literal, variable, or expression that returns a Boolean value, which can be True or False. Remember these two things:

If the value of the element next to the "if" keyword is equal to True, the program will process the statements within the "if" block.

If the value of the element next to the "if" keyword is equal to False, the program will skip or ignore the statements within the if block.

Else Statements

Else statements are used in conjunction with "if" statements. They are used to perform alternative statements if the preceding "if" statement returns False.

In the previous example, if the userAge is equal or greater than 18, the expression in the "if" statement will return False. And since the expression returns False on the "if" statement, the statements in the else statement will be executed.

On the other hand, if the userAge is less than 18, the expression in the "if" statement will return True. When that happens, the statements within the "if" statement will be executed while those in the else statement will be ignored. Mind you, an else statement has to be preceded by an "if" statement. If there is none, the program will return an error. Also, you can put an else statement after another else statement as long as it precedes an "if" statement.

In Summary

If the "if" statement returns True, the program will skip the else statement that follows.

If the "if" statement returns False, the program will process the else statement code block.

Code Blocks

Earlier in this book, the definition of a code block was discussed. Just to jog your memory, code blocks are simply groups of statements or declarations that follow **if** and **else** statements.

Creating code blocks is an excellent way to manage your code and make it efficient. In the coming chapters, you will mostly be working with statements and scenarios that will keep you working on code blocks.

Aside from that, you will learn about the variable scope as you progress. For now, you will mostly be creating code blocks "for" loops.

Loops

Loops are an essential part of programming. Every program that you use and see use loops.

Loops are blocks of statements that are executed repeatedly until a condition is met. It also starts when a condition is satisfied.

By the way, did you know that your monitor refreshes the image itself 60 times a second? Refresh means displaying a new image. The computer itself has a looping program that creates a new image on the screen.

You may not create a program with a complex loop to handle the display, but you will definitely use one in one of your programs.

A good example is a small snippet of a program that requires the user to login using a password.

For example:

>>> password = "secret"

>>> userInput = ""

>>> while (userInput != password):

　userInput = input()

_

This example will ask for user input. On the text cursor, you need to type the password and then press the Enter key. The program will keep on asking for user input until you type the word secret.

While

Loops are easy to code. All you need is the correct keyword, a conditional value, and statements you want to execute repeatedly.

One of the keywords that you can use to loop is **while**. While is like an "if" statement. If its condition is met or returns True, it will start the loop. Once the program executes the last statement in the code block, it will recheck the while statement and condition again. If the condition still returns True, the code block will be executed again. If the condition returns False, the code block will be ignored, and the program will execute the next line of code. For example >>> i = 1

```
>>> while i < 6:
 print(i)
 i += 1
1
2
3
4
5
>>> _
```

For Loop

While the **while** loop statement loops until the condition returns false, the "for" loop statement will loop at a set number of times depending on a string, tuple, or list. For example: >>> carBrands = ["Toyota", "Volvo", "Mitsubishi", "Volkswagen"]

```
>>> for brands in carBrands:
 print(brands)
Toyota
Volvo
```

Mitsubishi

Volkswagen

>>> _

Break

Break is a keyword that stops a loop. Here is one of the previous examples combined with break.

For example:

>>> password = "secret"

>>> userInput = ""

>>> while (userInput != password):

　userInput = input()

　break

　print("This will not get printed.")

Wrongpassword

>>> _

As you can see here, the while loop did not execute the print keyword and did not loop again after input was provided since the break keyword came after the input assignment.

The break keyword allows you to have better control of your loops. For example, if you want to loop a code block in a set amount of times without using sequences, you can use while and break.

```
>>> x = 0

>>> while (True):

 x += 1

 print(x)

 if (x == 5):

  break

1

2

3

4

5

>>> _
```

Using a counter, variable x (any variable will do of course) with an integer that increments every loop in this case, condition and break is common practice in programming. In most programming languages, counters are even integrated in loop statements. Here is a "for" loop with a counter in JavaScript.

for(i = 0; i < 10; i++) {

  alert(i);

}

This script will loop ten times. On one line, the counter variable is declared, assigned an initial value, a conditional expression was set, and the increments for the counter are already coded.

Infinite Loop

You should be always aware of the greatest problem with coding loops: infinity loops. Infinity loops are loops that never stop. And since they never stop, they can easily make your program become unresponsive, crash, or hog all your computer's resources. Here is an example similar to the previous one but without the counter and the usage of break.

\>>> while (True):

  print("This will never end until you close the program")

This will never end until you close the program

This will never end until you close the program

This will never end until you close the program

Whenever possible, always include a counter and break statement in your loops. Doing this will prevent your program from having infinite loops.

Continue

The continue keyword is like a soft version of break. Instead of breaking out from the whole loop, "continue" just breaks away from one loop and directly goes back to the loop statement. For example:

\>>> password = "secret"

\>>> userInput = ""

\>>> while (userInput != password):

userInput = input()

continue

print("This will not get printed.")

Wrongpassword

Test

secret

\>>> _

When this example was used on the break keyword, the program only asks for user input once regardless of anything you enter and it ends the loop if you enter anything. This version, on the other hand, will still persist on asking input until you put in the right password. However, it will always skip on the print statement and always go back directly to the while statement.

Here is a practical application to make it easier to know the purpose of the continue statement.

\>>> carBrands = ["Toyota", "Volvo", "Mitsubishi", "Volkswagen"]

\>>> for brands in carBrands:

if (brands == "Volvo"):

continue

print("I have a " + brands)

I have a Toyota

I have a Mitsubishi

I have a Volkswagen

\>>> _

When you are parsing or looping a sequence, there are items that you do not want to process. You can skip the ones you do not want to process by using a continue statement. In the above example, the program did not print "I have a Volvo" because it hit **continue** when a Volvo was selected. This caused it to go back and process the next car brand on the list.

Error Handling

Sometimes, errors happen during the program. This might be caused by a bad code or bad user input. Most of the time, it is the former.

Python immediately ends a program whenever errors are encountered. However, what if you want the show to continue despite these errors?

You might want to know what happens with the other code you have written after the line that produced the error. You want to know if they are also problematic. That is when error handling is useful.

Error handling is a programming process wherein you assume control of the program's errors from Python. Instead of just letting Python close your program, performing error handling can let you run code and continue with the program if an error is encountered.

Try and Except

One of the ways to handle errors is to use the keywords **try** and **except**. Try is like if. However, instead of testing a literal, variable, or expression's truth value, try only tests if the code block under it will generate an error.

"Except" statements work together with "try" statements. The purpose of except is to execute a code block when the code within the try statement returns an error. If you omit except and only use try, you will get an error. For example: >>> try:

a = 1

b = "a"

c = a + b

except:

print("There is an error on the try code block.")

There is an error on the try code block.

>>> _

In the above example, the try code block "tried" to add an integer and a string. Using the (+) operator like that will confuse Python. After all, the behavior of the (+) operator depends on the data type you use with it. If you use numbers, it will act as an addition operator. If you use strings, it will act as a concatenate operator. Normally, without the try statement, this will happen if you add an integer and a string:

>>> a = 1

>>> b = "a"

>>> c = a + b

Variable Styling

Here are a few quick reminders from Python's style guide (PEP 8).

As much as possible, sparingly use global variables. And when you truly need one, just make sure that the set of global variables you will use is for a single module only.

Again, do not use the lower case l, uppercase O, or the uppercase I for single-letter variables. As you can see right now, it is difficult to differentiate l, I, and 1 and O and 0 from each other.

Practice Exercise

For this chapter, create a choose-your-adventure program. The program should provide users with two options. It must also have at least five choices and have at least two different endings.

You must also use dictionaries to create dialogues.

Here is an example:

creepometer = 1

prompt = "\nType 1 or 2 then press enter...\n\n ::> "

clearScreen = ("\n" * 25)

scenario = [

 "You see your crush at the other side of the road on your way to school.",

 "You notice that her handkerchief fell on the ground.",

 "You heard a ring. She reached on to her pocket to get her phone and stopped.",

 "Both of you reached the pedestrian crossing, but its currently red light.",

"You got her attention now and you instinctively grabbed your phone."

]

choice1 = [

"Follow her using your eyes and cross when you reach the intersection.",

"Pick it up and give it to her.",

"Walk pass her.",

"Smile and wave at her.",

"Ask for her number."

]

choice2 = [

"Cross the road and jaywalk, so you will be behind her.",

"Pick it up and keep it for yourself.",

"Stop and pretend you are tying your shoes.",

"Tap her shoulders.",

"Take a picture of her using your phone."

]

```python
result1 = [

"A car honked at you and she noticed you. She walked a bit faster.",

"You called her and you gave her the handkerchief.",

"She noticed you as you walked pass her, but she focused on the call she got.",

"She smiled and waved back.",

"She started to think about it."

]

result2 = [

"You walked casually and crossed the pedestrian lane.",

"You stashed away her handkerchief on your pocket.",

"She noticed you and her rightbrow rose.",

"She turned towards you.",

"Her eyes suddenly become bloodshot red."

]

ending1 = [

"She grabbed her phone, and typed some numbers.",

"You became giddy.",
```

"After a second, she showed you her phone.",

"Her number was on the screen.",

"You quickly fiddled with your phone and typed in her digits.",

"She walked away towards the school gate."

]

ending2 = [

"She politely turned down your request.",

"She walked away towards the school gate.",

"She looked back at you for a moment.",

"Your eyes met for a moment.",

"Then she turned away.",

"There is hope for you, you thought."

]

ending3 = [

"Her right hand moved and the next thing you saw was the sky.",

"Your life flashed in front of you.",

"Her scream brought you back to reality.",

"Your left cheek was scorched hot as the pain radiate from it.",

"You then asked yourself why.",

"That was the last time you saw her."

]

instructions = [

"Here are the instructions on how to play this game.",

"1. To play and complete this game, you must enter your choices when asked.",

"2. Press enter to proceed with the next dialog.",

"3. The choices you make changes the ending of the game.",

"Press enter whenever you are ready."

]

print(clearScreen)

for i in range(len(instructions)):

  print(instructions[i])

input()

print(clearScreen)

for i in range(len(scenario)):

```
input(scenario[i])

print("1. " + choice1[i])

print("2. " + choice2[i])

answer = ""

while (True):

answer = input(prompt)

if(answer == "1" or answer == "2"):

break

print("\n")

if(answer == "1"):

input(result1[i])

creepometer -= 1

else:

input(result2[i])

creepometer += 1

if(creepometer < 0):

for i in range(len(ending1)):
```

```
    input(ending1[i])

if(creepometer == 0):

    for i in range(len(ending2)):

        input(ending2[i])

if(creepometer > 0):

    for i in range(len(ending3)):

        input(ending3[i])

input("Thank you for playing the game!")
```

By the way, the clearScreen variable contains multiplied \n (new line) characters. Printing numerous new lines can push the previous lines upwards, which basically clears the screen.

You can make the input() function add a prompt by passing a string inside its parentheses. Also, it is an excellent way to "pause" the program and wait for users to press enter to continue.

Have fun!

CHAPTER 9:

Strings

Values-How to Access Them

At the point in time that Python is not supporting a specific character type, it is then treated like a string that may or may not have a substring that is attached to it.

When you want to get into the substring, you are going to need to use a set of brackets ([ ]) in order to ensure that the string is cut along the correct index so that the substring is created properly.

Example

#!/usr/bin/python

Variable 1 = 'This is a string I created'

Variable 2 = "I can type anything I want"

Print "Variable 1 [1]: " , variable 1 [1]

Print "Variable 2 [3:6]: " , variable 2 [3:6]

Output

Variable 1 [1]: h

Variable 2 [3: 6]: an t

It may not make much sense by the example, but it is going to be more helpful when you are trying to create your substrings.

It is also going to make more sense after you have made the string yourself because it is going to be your own creation.

Updating

When you are going through your strings and notice that something needs to be changed, you are going to be able to do this without having to rewrite the entire string over.

Any new value that you want to move or reassign is going to be placed into a new string or it is going to be changed to a different value of your choosing.

Example

#!/usr/bin/python

Variable 1 = 'This is what the world works.'

Print "update string : - ", variable 1 [: 3] + 'how'

Output

Update string : - 'This is how the world works.'

Escape Characters

When you use an escape character, it is going to be interpreted by Python.

Escaped characters are going to use single or double-quotes.

Before you can use an escaped character, you are going to need to use a backslash so that Python knows what you are trying to do.

\a – a bell or an alert

\b – backspaces

\ cx or \C-x – Control -x

\ e – escape

\f – formfeed

\ M-\C-x – The Meta Control x method

\n – a new line will be created

\ nnn-n is going to be 0.7 for the octal notation

\r – the carriage will be returned

\ s – a space will be inserted

\ t – a tab is going to be inserted

\ v – a vertical tab will be used

\ x – you can insert the character x

\ xnn – the hexadecimal notation is going to be 0.9

Special Operators

When you want something to happen with the string that you have created but there is not a specific function for it, then you are going to need to do a special operation so that you can complete the expression that you are wanting to do.

Concatenation

The value is going to be added despite where it falls on the operator (+).

Repetition

A new string will be created but it is going to be the same string that has already been made. **(*) Slice**

The string is going to be cut at a specific character depending on what index is indicated ([ ]).

Range Slice

The characters are going to be chosen between the set range ([:]).

Membership

The condition is going to be returned as true but only if the character is found inside of the selected string (in).

Membership

This is the same thing as the previous one, except the character, is not going to be found inside of the string that you have selected (not in).

Raw String

The escaped characters are going to be suppressed. You are going to write it just like you would if you were creating a now string, except you are going to use the letter r with a set of quotation marks.

(r / R) Format

The string is going to be formatted (%).

Formatting a String

To format a string, you are going to use the percent sign.

This function is only going to work when you are formatting strings.

Example

#!/usr/bin/python

Print "his name was %s and he was a %d." % (Tim, bodybuilder)

Output

"His name was Tim and he was a bodybuilder."

There are other symbols that you can use with the percent sign to get different results when you are formatting your string.

%c – a character needs to be inserted

%s – the string needs to be converted with the use of the str() function before it can be formatted %i and %d – a decimal will be entered, and it will be signed.

%u – the decimal put in will be unsigned.

%o – the integer is going to be octal

%x – the hexadecimal integer will be lowercased

%X – the hexadecimal integers will be uppercased

%e – with a lowercase e, the exponential notation is going to be inserted

%E – with an uppercase E, the exponential notation will be inserted.

%f – floating point numbers are going to be real numbers

%g – the same thing, but a shorter version for %f and %e

%G – the shorter version of %f and %E

There are other special characters that can be used with Python strings that are not made into functions.

When your argument has a specific argument for its precision (*).

The justification will be to the left (-).

The sign is going to be displayed (+).

There is going to be a space that is left before the positive number (<sp>).

The octal will be zero or the hexadecimal will be 0x. The version of x that you use will determine which it is (#) There will be a set number of pads to the left using zeros (0).

A single literal will be placed in the equation (%).

The variables will be mapped based on the arguments in the dictionary (var).

There is going to be a minimum number of digits that are going to be displayed after the decimal point and there will be a set width. (m. n.).

Triple Quotes

Special characters are going to come with special rules and that includes using triple quotes.

Some other special characters that are going to use triple quotes are tabs as well as new lines. Triple quotes are going to use either three single quotes or three sets of double-quotes.

Example

#!/usr/bin/python

Para_str = """I am going to type for a long time so that it makes it seem like it is going to be a very long piece of script. In this script, the characters that I am going to use are \t which is going to insert a tab wherever I place that character. New lines can also be added to this script by using [\ n] or by typing in a new line so that the variable shows up where it needs to. Python is going to execute the code above just like any other piece of code that it encounters. When it does, it is going to not insert the functions that are inside of the script. Instead, it is going to put the correct object in. So, where there is a tab, a tab is going to be placed in it.

A backslash in a raw string is not going to be treated like it is a special character. Instead, whenever you put a character, into a raw string, it is going to be produced like it appears.

Example

#! *usr* bin/ python

Print 'D:\\ anywhere'

Output

D: \ anywhere

But, when you put in an expression such as an r expression it is going to change the output.

Example

#!/usr/bin/python

Print 'D:\\ anywhere'

Output

D:\\ anywhere

Unicode

A string with Python is going to have an ASCII that is made up of 8 bits.

The Unicode strings will have 16 bits instead of 8.

Having more bits is going to allow for more characters and special characters to be used.

Example

#!/usr/bin/python

Print u 'I am lost!'

Output

I am lost!

The 'u' that you see before the printout is going to be what tells Python that you are using Unicode instead of a different function.

As we are working on some of the strings that we want to handle, you will find that there are a lot of options that are going to be available for you to use at any time. These strings come with a lot of functions and will help you to add more options and features to some of the codes that you are trying to write.

Some of the different functions that are going to work well when you design strings in Python will include:

Capitalize():

This one is going to take the first letter of the string and capitalize it for you.

Center(width, char):

This is going to return to you a string that is at least the specified width, and then it will be created by padding the string with the character.

Count(str):

This is going to return the number of times that a particular string is contained in another string.

Find(str):

This is going to return the index number of the substring in the string.

Isalpha():

This is going to check if all the characters of a string are alphabetic characters.

Isdigit():

This part is going to check whether the string contains just numbers or digits or if there is a mixture.

Islower):

This function is going to take a look to see if the string you are checking has all lower case characters.

Len():

This is going to let you know the length of the string

Isupper():

This one is going to check to see if all the characters in the string are upper case.

Lower():

This will give you a return that has all the string in lower case letters.

Replace():

This is going to take the string that you have and replace it with a new string.

Upper():

This is going to return the string in upper case.

Split():

This is going to split up the string based on the split character.

CHAPTER 10:

Functions

Functions are code blocks that are given an identifier. This identifier can be used to call the function. Calling a function makes the program execute the function regardless of where it is located within the code.

To create a function, you need to use the "def" keyword. Def basically defines, and when you use it to create a function, you can call it as defining a function. For example:

>>> def doSomething():

 print("Hello functioning world!")

>>> doSomething()

Hello functioning world!

>>> _

Creating and calling a function is easy. The primary purpose of a function is to allow you to organize, simplify, and modularize your code. Whenever you have a set of code that you will need to execute in sequence from time to time, defining a function for that set of code will save you time and space in your program. Instead of repeatedly typing code or even copy-pasting, you simply define a function.

We began with almost no prior knowledge about Python except for a clue that it was some kind of programming language that is in great demand these days. Now, look at you; creating simple programs, executing codes, and fixing small-scale problems on your own. Not bad at all! However, learning always comes to a point where things can get rather trickier.

In quite a similar fashion, Functions are docile looking things; you call them when you need to get something done. But did you know that these functions have so much going on at the back? Imagine every function as a mini-program. It is also written by programmers like us to carry out specific things without having us write lines and lines of codes. You only do it once, save it as a function, and then just call the function where it is applicable or needed.

The time has come for us to dive into a complex world of functions where we don't just learn how to use them effectively, but we also look into what goes on behind these functions, and how we can come up with our very own personalized function. This will be slightly challenging, but I promise, there are more references that you will enjoy keeping the momentum going.

Functions are like containers that store lines and lines of codes within themselves, just like a variable that contains one specific value. There are two types of functions we get to deal with within Python. The first ones are built-in or predefined, the other is custom-made or user-created functions.

Either way, each function has a specific task that it can carry out. The code that is written before creating any function is what gives that function an identity and a task. Now, the function knows what it needs to do whenever it is called in.

When we began our journey, we wrote "I made it!" on the console as our first program? We used our first function there as well: the print() function. Functions are generally identified by parentheses that follow the name of the function. Within these parentheses, we pass arguments called parameters. Some functions accept a certain kind of parenthesis while others accept different ones.

Let us look a little deeper and see how functions greatly help us reduce our work and better organize our codes. Imagine, we have a program that runs during live streaming of an event. The purpose of the program is to provide our users with a customized greeting. Imagine just how many times you would need to write the same code again and again if there were quite a few users who decide to join your stream. With functions, you can cut down on your own work easily.

In order for us to create a function, we first need to 'define' the same. That is where a keyword called 'def' comes along. When you start typing 'def' Python immediately knows you are about to define a function.

You will see the color of the three letters change to orange (if using PyCharm as your IDE). That is another sign of confirmation that Python knows what you are about to do.

def say_hi():

Here, say_hi is the name I have decided to go with, you can choose any that you prefer. Remember, keep your name descriptive so that it is understandable and easy to read for anyone. After you have named your function, follow it up with parentheses. Lastly, add the friendly old colon to let Python know we are about to add a block of code. Press enter to start a new indented line. Now, we shall print out two statements for every user who will join the stream.

print("Hello there!")

print('Welcome to My Live Stream!')

After this, give two lines of space to take away those wiggly lines that appear the minute you start typing something else. Now, to have this printed out easily, just call the function by typing its name and run the program. In our case, it would be:

say_hi()

Output:

Hello there!

Welcome to My Live Stream!

See how easily this can work for us in the future? We do not have to repeat this over and over again. Let's make this function a little more interesting by giving it a parameter. Right at the top line, where it says "def say_hi()"? Let us add a parameter here. Type in the word 'name' as a parameter within the parenthesis. Now, the word should be greyed out to confirm that Python has understood the same as a parameter.

Now, you can use this to your advantage and further personalize the greetings to something like this:

def say_hi(name):

print(f"Hello there, {user}!")

print('Welcome to My Live Stream!')

user = input("Please enter your name to begin: ")

say_hi(user)

The output would now ask the user regarding their name. This will then be stored into a variable called user. Since this is a string value, say_hi() should be able to accept this easily. Bypassing 'user' as an argument, we get this as an output:

Please enter your name to begin: Johnny

Hello there, Johnny!

Welcome to My Live Stream!

Now that's more like it! Personalized to perfection. We can add as many lines as we want, the function will continue to update itself and provide greetings to various users with different names.

There may be times where you may need more than just the user's first name. You might want to inquire about the last name of the user as well. To add to that, add this to the first line and follow the same accordingly:

def say_hi(first_name, last_name):

  print(f"Hello there, {first_name} {last_name}!")

  print('Welcome to My Live Stream!')

first_name = input("Enter your first name: ")

last_name = input("Enter your last name: ")

say_hi(first_name, last_name)

Now, the program will begin by asking the user for their first name, followed by the last name. Once that is sorted, the program will provide a personalized greeting with both the first and last names.

However, these are positional arguments, meaning that each value you input is in order. If you were to change the positions of the names for John Doe, Doe will become the first name and John will become the last name. You may wish to remain a little careful about that.

Hopefully, now you have a good idea of what functions are and how you can access and create them. Now, we will jump towards a more complex front of 'return' statements.

"Wait! There's more?" Well, I could have explained, but back then, when we were discussing statements, you may not have understood it completely. Since we have covered all the bases, it is appropriate enough for us to see exactly what these are and how these gel along with functions.

Return Statement

Return statements are useful when you wish to create functions whose sole job is to return some values. These could be for users or for programmers alike. It is a lot easier if we do this instead of talk about theories, so let's jump back to our PyCharm and create another function.

Let us start by defining a function called 'cube' which will basically multiply the number by itself three times. However, since we want Python to return a value, we will use the following code:

def cube(number):

return number ***number*** number

By typing 'return' you are informing Python that you wish for it to return a value to you that can later be stored in a variable or used elsewhere. It is pretty much like the input() function where a user enters something and it gets returned to us.

def cube(number):

return number *number* number

number = int(input("Enter the number: "))

print(cube(number))

Go ahead and try out the code to see how it works. It is not necessary that we define functions such as these. You can create your own complex functions that convert kilos into pounds, miles into kilometers, or even carry out far greater and more complex jobs. The only limit is your imagination. The more you practice, the more you explore. With that said, it is time to say goodbye to the world of functions and head into the advanced territories of Python. By now, you already have all you need to know to start writing your own codes.

How to Define and Call Function?

To start, we need to take a look at how we are able to define our own functions in this language. The function in Python is going to be defined when we use the statement of "def" and then follow it with a function name and some parentheses in place as well. This lets the compiler know that you are defining a function, and which function you would like to define at this time as well. There are going to be a few rules in place when it comes to defining one of these functions though, and it is important to do these in the proper manner to ensure your code acts in the way that you would like. Some of the Python rules that we need to follow for defining these functions will include:

1. Any of the arguments or input parameters that you would like to use have to be placed within the parentheses so that the compiler knows what is going on.

2. The function first statement is something that can be an optional statement something like a documentation string that goes with your function if needed.

3. The code that is found within all of the functions that we are working with needs to start out with a colon, and then we need to indent it as well.

4. The statement return that we get, or the expression, will need to exit a function at this time. We can then have the option of passing back a value to the caller. A return statement that doesn't have an argument with it is going to give us the same return as None.

Before we get too familiar with some of the work that can be done with these Python functions, we need to take some time to understand the rules of indentation when we are declaring these functions in Python. The same kinds of rules are going to be applicable to some of the other elements of Python as well, such as declaring conditions, variables, and loops, so learning how this work can be important here.

You will find that Python is going to follow a particular style when it comes to indentation to help define the code because the functions in this language are not going to have any explicit begin or end like the curly braces in order languages to help indicate the start and the stop for that function. This is why we are going to rely on the indentation instead. When we work with the proper kind of indentation here, we are able to really see some good results and ensures that the compiler is going to know when the function is being used.

Parameters

Parameters Require Arguments

You cannot call a function with parameters without an argument. If you do, you will receive an error. For example:

>>> def sampFunc(x):

 print(x)

>>> sampFunc()

Traceback (most recent call last):

 File "<stdin>", line 1, in <module>

TypeError: y() missing 1 required positional argument: 'x'

>>> _

Multiple Parameters

You can assign two or more parameters in a function. For example:

```
>>> def simpOp(x, y):

 z = x + y

 print(z)

>>> simpOp(1, 2)

3

>>> _
```

Return statement

Returning Value

The return keyword makes a function return a value. For a simpler explanation, it makes the function be used as a variable that has an assigned or processed value. For example:

```
>>> def concat(string1, string2):

 return string1 + string2

>>> x = concat("Text1", "Text2")

>>> x

'Text1Text2'
```

```
>>> _
```

A function can return a value even if it does not have parameters. For example:

```
>>> def piString():
  return "3.14159265359"
>>> x = piString()
>>> x
'3.14159265359'
>>> _
```

As you can see, using the keyword method makes it simpler for you to retrieve a value from a function without relying on global variables. Return allows you to make a clean and efficient code.

Lambada Function

Using an anonymous function is a convenient way to write one-line functions that require arguments and return a value. It uses the keyword lambda. Despite having a purpose of being a one liner, it can have numerous parameters. For example:

```
>>> average = lambda x, y, z: (x + y + z) / 3
>>> x = average(10, 20, 30)
```

```
>>> x

20.0

>>> average(12, 51, 231)

98.0

>>> _
```

Global Variables

Global variables utilized in Python are a multipurpose variable used in any part of the world while anywhere. The variable used can operate in your program or module while in any part of the globe henceforth using values whenever you travel as a programmer. Global variables are useful for programmers to create their programs while moving from one location to another. Some of the benefits include variables that are used across function or module as well as it does not require re-declarations for performance.

When compared to local variables, global variables have an 'f' scope and assigned value 101, displayed as 'f=101', printed as an output. For example, when you re-declare a variable as a global variable in a given function, change it within the role and print it outside the task. The variable would provide a third-party outcome useful globally. Therefore, global variables are found outside functions indicating that not all variables are readily accessed from anywhere globally.

As a beginner, it is crucial to understand the difference between global and local variables to develop the necessary variables suitable for your programs.

Local Variables

Unlike global variables, local variables are used locally, declared within a Python function or module, and utilized solely in a specific program or Python module. When implemented outside particular modules or tasks, the Python interpreter will fail to recognize the units henceforth throwing an error message for undeclared values. Like global variables, local variables use the 'f' variable where it is declared to assume local scope and assigned 'I am learning Python' and then recognized as a local variable.

For example, when you declare the variable 'f' the second time, it changes to a new function and results in a local variable. As such, when you accept that variable in the inside function, the process will run without any problems. Whereas, when you print the value outside the function 'f', it results in a value assigned to it, which is outside the function, that is, a third print(f). In that case, local variables are used only in two surrounding environments of Python programming with those outside the function leading to failure of operations unless declared.

CHAPTER 11:

Dictionaries

Dictionaries are another data storage container similar to lists and tuples with one key difference. Dictionaries store key:value pairs.

Similar to lists, dictionaries are mutable and can contain mixed content including tuples (assuming the values are all of the same types), lists, strings, integers, etc.

The point of the key:value pair model is that every key stored within the dictionary is unique. In effect, the dictionary is designed to store the values of a list of unique items or properties.

Given that the following things are true:

1. When adding to or creating a dictionary, all entries must be in a key:value pair format. If either the key or the value is missing an exception will be given.

2. If a key:value pair is added to a dictionary and the key already exists in the dictionary, the old key and value will be discarded for the new entry.

3. Values can be updated only when a key is provided.

4. key:value pairs are always deleted together. There is no way to delete just the value or just the key.

5. Lists cannot be used as keys.

Here are some common operations on dictionaries.

Create an empty dictionary:

>>> x={}

Create a dictionary:

>>> x=dict(name='john', age=35, height=6)

>>> print(x)

{'name': 'john', 'age': 35, 'height': 6}

Test if a key exists:

>>> if 'age' in x.keys():

print('found it!')

found it!

Get a value for a key:

>>> y=x['height']

>>> print('Height = ' + str(y))

Height = 6

List all keys:

```
>>> for y in x.keys():
 print(y)
name
 age
 height
```

Print keys and values:

```
>>> for y in x.keys():
 v=x[y]
 if type(v)!='str':
  v=str(v)
 print(y + '=' + v)
name=john
 age=35
 height=6
```

Update a value for a key:

```
>>> x['age']=40
>>> print(x)
 {'name': 'john', 'age': 40, 'height': 6}
```

Delete a key:value pair:

>>> del(x['height'])

>>> print(x)

{'name': 'john', 'age': 40}

Add a Key:Value pair:

>>> x['height']=6

>>> print(x)

{'name': 'john', 'age': 40, 'height': 6}

Dumping all keys or values into a list:

>>> list(x.keys()) # keys as a list

['name', 'age', 'height']

>>> list(x.values()) # values as a list

['john', 40, 6]

CHAPTER 12:

Python Operators

Operators are symbols that signify the execution of a specific process. Python provides several types of operators:

Arithmetic Operators

Python supports 7 arithmetic operators:

Addition +

Subtraction -

Multiplication *

Division /

Exponent **

Modulos %

Floor Division //

Subtraction (-)

The subtraction operators subtract the value of the right operand from that of the left operand.

>>>13 – 4

9

Multiplication (*)

The multiplication operator multiplies the left and right operands.

>>>12 * 3

36

Division (/)

Exponent (**)

The exponent operator raises the base number to the power signified by the number after the operator.

>>> 4**2

16

Modulos (%)

The modulos operator returns the remainder after performing a division operation of the left operand with the right operand.

>>> 20 % 6

2

Floor Division (//)

The floor division operator performs a division operation, drops the fractional part, and returns the quotient as a whole number.

>>> 20 // 3

6

Assignment Operators

Assignment operators are used to assign values to variables.

= Operator

The = operator assigns the right operand to the left operand.

Examples:

x = 35

a = b

vowel_list = ['a', 'e', 'i', 'o', 'u']

new_dict = {'Name':'Brandon Smart', 'Age':25, 'Employment Status: ':'Regular'}

x = 2 * 4

Python supports multiple assignments in a single statement:

a, b, c = "Polar bear", 12, 5.5

Python likewise allows the assignment of one value to several variables in a single statement:

a = b = c = "high cube"

num = item = sum = 27

add and +=

The 'add and' operator adds the value of the left and right operands and assigns the total to the left operand.

x += 4

y += x

subtract and -=

The 'subtract and' operator subtracts the value of the right operand from that of the left and assigns the difference to the left number.

x -= y

a -= 4

multiply and *=

The 'multiply and' operator multiplies the left and right operands and assigns the product to the left operand.

x *= z

a *= 4

divide and /=

x /= c

y /= 4

modulos and %=

The 'modulos' and operator divides the value of the left operand with the right

operand then assigns the remainder to the left.

x %= a

x %= 3

floor division and //=

The 'floor division and' operator performs a floor division of the left operand by the right operand and assigns the result to the left operand.

x //= a

x //= 2

Relational or Comparison Operators

Relational operators evaluate a comparative expression and return either True or False.

Python provides the following relational operators:

Operator	Meaning
>	is greater than
<	is less than
==	is equal to
!=	is not equal to
>=	is greater than or equal to
<=	is less than or equal to

Examples:

>>> 25 == 5*2*3

False

>>> 3*4 <= 3*2*3

True

>>> 45 >= (15*5)

False

>>> 36 != 3**2*4

False

```
>>> (12*3) > 30
```

True

```
>>> 30 < (3*15)
```

True

Logical Operators

Python provides three kinds of logical operators:

or

and

not

Python evaluates expressi0ns with logical operators by applying the following tests:

x = first argument, y = second argument

x or y if x is true, it returns True. If x is false, it evaluates y and returns the result as either True or False. In other words, only one argument needs to be True for the expression to return True.

Examples:

```
>>> (24>9) or (12<9) #The first argument is true.
```

True

>>> (7 > 14) or (5 < 15) #The second argument is true.

True

x and y If x is true, it evaluates y. If y is true, it returns True. If y is false, it returns False. If x is false, it returns False. In other words, both arguments should be true in order for the operation to return True.

Examples:

>>> (12>23) and (25>10) # The first argument is false.

False

>>> (15 == 3*5) and (15 < 3**4) #Both arguments are true.

True

not x If x is true, it returns False. Otherwise, it returns True.

Examples:

>>> not (5*4 > 20)

True

>>> not (15 > 4**3)

True

Identity Operators

"is" returns True if the specified variables refer to the same object, and returns False if otherwise.

"is not" returns False if the specified variables refer to the same object or memory location and True if otherwise.

Examples:

\>>> a = 12

\>>> b = 12

\>>> a is b

True

The variables a and b contain integers of similar value which makes them identical and equal.

\>>> x = 'immutable'

\>>> y = 'immutable'

\>>> x is not y

False

The variables x and hold the same string and data type and are identical and equal. Hence, the 'is not' identity operator returned False.

>>> list_a = [12, 24, 36]

>>> list_b = [12, 24, 36]

>>> list_a is list_b

False

The variables list_a and list_b contains the same elements and are equal. However, they are not identical because lists are mutable and are thus saved separately in memory. Hence, the id operator 'is' returned False.

CHAPTER 13:

Working with Files

Programs are made with input and output in mind.

You input data to the program, the program processes the input, and it ultimately provides you with output.

For example, a calculator will take in numbers and operations you want.

It will then process the operation you wanted.

And then, it will display the result to you as its output.

There are multiple ways for a program to receive input and to produce output.

One of those ways is to read and write data on files.

To start learning how to work with files, you need to learn the open() function.

The open() function has one *required* parameter and two *optional* parameters.

The first and required parameter is the file name.

The second parameter is the access mode.

And the third parameter is buffering or buffer size.

The filename parameter requires string data.

The access mode requires string data, but there is a set of string values that you can use and is defaulted to "r".

The buffer size parameter requires an integer and is defaulted to 0.

To practice using the open() function, create a file with the name sampleFile.txt inside your Python directory.

Try this sample code:

```
>>> file1 = open("sampleFile.txt")

>>> _
```

Note that the file function returns a file object.

The statement in the example assigns the file object to variable file1.

The file object has multiple attributes, and three of them are:

Name

This contains the name of the file.

Mode

This contains the access mode you used to access the file.

Closed

This returns False if the file has been opened, and True if the file is closed. When you use the open() function, the file is set to open.

Now, access those attributes.

\>>> file1 = open("sampleFile.txt")

\>>> file1.name

'sampleFile.txt'

\>>> file1.mode

'r'

\>>> file1.closed

False

\>>> _

Whenever you are finished with a file, close them using the close() method.

\>>> file1 = open("sampleFile.txt")

\>>> file1.closed

False

\>>> file1.close()

```
>>> file1.closed
```

True

```
>>> _
```

Remember that closing the file does not delete the variable or object.

To reopen the file, just open and reassign the file object.

For example:

```
>>> file1 = open("sampleFile.txt")
```

```
>>> file1.close()
```

```
>>> file1 = open(file1.name)
```

```
>>> file1.closed
```

False

```
>>> _
```

Reading from a File

Before proceeding, open the sampleFile.txt in your text editor.

Type "Hello World" in it and save.

Go back to Python.

To read the contents of the file, use the read() method.

For example:

\>\>\> file1 = open("sampleFile.txt")

\>\>\> file1.read()

'Hello World'

\>\>\> _

File Pointer

Whenever you access a file, Python sets the file pointer.

The file pointer is like your word processor's cursor.

Any operation on the file starts at where the file pointer is.

When you open a file, and when it is set to the default access mode, which is "r" (read-only), the file pointer is set at the beginning of the file.

To know the current position of the file pointer, you can use the tell() method.

For example:

\>\>\> file1 = open("sampleFile.txt")

\>\>\> file1.tell()

0

>>> _

Most of the actions you perform on the file move the file pointer.

For example:

>>> file1 = open("sampleFile.txt")

>>> file1.tell()

0

>>> file1.read()

'Hello World'

>>> file1.tell()

11

>>> file1.read()

''

>>> _

To move the file pointer to a position you desire, you can use the seek() function.

For example:

>>> file1 = open("sampleFile.txt")

>>> file1.tell()

0

>>> file1.read()

'Hello World'

>>> file1.tell()

11

>>> file1.seek(0)

0

>>> file1.read()

'Hello World'

>>> file1.seek(1)

1

>>> file1.read()

'ello World'

>>> _

The seek() method has two parameters.

The first is offset, which sets the pointer's position depending on the second parameter.

Also, an argument for this parameter is required.

The second parameter is optional.

It is for whence, which dictates where the "seek" will start.

It is set to 0 by default.

If set to 0, Python will set the pointer's position to the offset argument.

If set to 1, Python will set the pointer's position relative or in addition to the current position of the pointer.

If set to 2, Python will set the pointer's position relative or in addition to the files end.

Note that the last two options require the access mode to have binary access.

If the access mode does not have binary access, the last two options will be useful to determine the current position of the pointer [seek(0, 1)] and the position at the end of the file [seek(0, 2)].

For example:

>>> file1 = open("sampleFile.txt")

>>> file1.tell()

0

>>> file1.seek(1)

1

```
>>> file1.seek(0, 1)
0
>>> file1.seek(0, 2)
11
>>> _
```

File Access Modes

To write to a file, you will need to know more about file access modes in Python.

There are three types of file operations: reading, writing and appending.

Reading allows you to access and copy any part of the file's content.

Writing allows you to overwrite a file's contents and create a new one.

Appending allows you to write on the file while keeping the other content intact.

There are two types of file access modes: string and binary.

String access allows you to access a file's content as if you are opening a text file.

Binary access allows you to access a file in its rawest form: binary.

In your sample file, accessing it using string access allows you to read the line "Hello World".

Accessing the file using binary access will let you read "Hello World" in binary, which will be b'Hello World'.

For example:

\>>> x = open("sampleFile.txt", "rb")

\>>> x.read()

b'Hello World'

\>>> _

String access is useful for editing text files.

Binary access is useful for anything else, like pictures, compressed files, and executables. In this book, you will only be taught how to handle text files.

There are multiple values that you can enter in the file access mode parameter of the open() function.

But you do not need to memorize the combination.

You just need to know the letter combinations.

Each letter and symbol stands for an access mode and operation.

For example:

r = read-only—file pointer placed at the beginning

r+ = read and write

a = append—file pointer placed at the end

a+ = read and append

w = overwrite/create—file pointer set to 0 since you create the file

w+ = read and overwrite/create

b = binary

By default, file access mode is set to string.

You need to add b to allow binary access.

For example: "rb".

Writing to a File

When writing to a file, you must always remember that Python overwrites and does not insert file.

For example:

>>> x = open("sampleFile.txt", "r+")

>>> x.read()

'Hello World'

```
>>> x.tell(0)

0

>>> x.write("text")

4

>>> x.tell()

4

>>> x.read()

'o World'

>>> x.seek(0)

0

>>> x.read()

'texto World'

>>> _
```

You might have expected that the resulting text will be "textHello World".

The write method of the file object replaces each character one by one, starting from the current position of the pointer.

Practice Exercise

For practice, you need to perform the following tasks:

Create a new file named test.txt.

Write the entire practice exercise instructions on the file.

Close the file and reopen it.

Read the file and set the cursor back to 0.

Close the file and open it using append access mode.

Add a rewritten version of these instructions at the end of the file.

Create a new file and put similar content to it by copying the contents of the test.txt file.

Summary

Working with files in Python is easy to understand but difficult to implement.

As you already saw, there are only a few things that you need to remember.

The hard part is when you are actually accessing the file.

Remember that the key things that you should master are the access modes and the management of the file pointer.

It is easy to get lost in a file that contains a thousand characters.

Aside from being versed in the file operations, you should also supplement your learning with the functions and methods of the str class in Python.

Most of the time, you will be dealing with strings if you need to work on a file.

Do not worry about binary yet.

That is a different beast altogether and you will only need to tame it when you are already adept at Python.

As a beginner, expect that you will not deal yet with binary files that often contain media information.

Anyway, the next lesson is an elaboration on the "try" and "except" statements.

You'll discover how to manage and handle errors and exceptions effectively.

Conclusion

Thank you for reading this book. The power of programming languages in our digital world cannot be underestimated. People are increasingly reliant on the modern conveniences of smart technology and that momentum will endure for a long time. With all the instructions provided in this book, you are now ready to start developing your own innovative smart tech ideas and turn it into a major tech startup company and guide mankind towards a smarter future.

It is by no means a comprehensive lesson on coding, but I hope that I have been able to give you the basics, enough for you to be able to move on and expand your learning.

The thing about computer programming is that your learning will never stop. Even if you think that you have the basics down pat if you don't use what you have learned regularly. Believe me when I say that you will soon forget it! Computer programming is evolving on an almost daily basis and it's up to you to keep up with everything that is going on. To that end, you would be well advised to join a few of the Python communities. You will find many of these on the internet and they are places where you can stay up to the minute with changes, where you can join in conversations, discuss code, and ask for help.

Eventually, you will be in a position of being able to help the newbies on the scene and it is then that you will realize just how far you have come.

Don't just read it once and think that you know it all because you don't. The human brain can only take in so much information in one go and it needs time to assimilate that information and store it away before the next influx. Trying to take in pages and pages of code and information will not serve you well and it isn't a case of being the quickest to read it. You can read as much as you like but, once your brain stops taking the information in, you anything else will be meaningless.

Take your time; do the exercises as many times as you need to until you know that you can write the answers AND understand the answers in your sleep. That is important – it is not enough to know the answers with Python programming. You have to be able to understand WHY the answer is such, the process that gets to that answer if you don't understand the code from start to finish you will never be able to understand the answers.

Python is a valuable programming language with a large array of uses. It is practical, efficient, and extremely easy to use. It will be a great asset and reference point for your future in programming. If you can think it, you can create it. Don't be afraid to try something new.

Remember that knowledge is useless without application. Learning how to program without actually programming will only waste the time you invested in this book.

This book is meant to help a beginner programmer learn Python. With this book, you should be able to write basic programs and even more complex programs with multiple objects.

We provided plenty of exercises for you to practice your programming. Becoming a Pythonista will require plenty of practice. Feel free to come up with your own exercises and practice as well.

Good luck and happy programming!

CPSIA information can be obtained
at www.ICGtesting.com
Printed in the USA
LVHW050235161120
671797LV00008B/255